# An Introduction to Combinatorial Analysis

# An Introduction to Combinatorial Analysis

JOHN RIORDAN

*The Rockefeller University*

*Former Member Technical Staff*

*Bell Telephone Laboratories, Inc.*

*Princeton University Press*

Princeton, New Jersey

Published by Princeton University Press, Princeton, N.J.
In the U.K.: Princeton University Press, Guildford, Surrey



LCC 80-337
ISBN 0-691-08262-6
ISBN 0-691-02365-4 pbk.

Originally published 1958 by John Wiley & Sons, Inc.
First Princeton University Press printing, 1980

*To E. T. Bell*

# Preface

COMBINATORIAL ANALYSIS, THOUGH A WELL-RECOGNIZED PART of mathematics, seems to have a poorly defined range and position. Leibniz, in his "ars combinatoria", was the originator of the subject, apparently in the sense of Netto (*Lehrbuch der Combinatorik,* Leipzig, 1901) as the consideration of the placing, ordering, or choice of a number of things together. This sense appears also in the title, *Choice and Chance* (W. A. Whitworth, fifth edition, London, 1901), of one of the few books in English on the subject. This superb title also suggests the close relation of the subject to the theory of probability. P. A. MacMahon, in the most ambitious treatise on the subject (*Combinatory Analysis,* London, vol. I, 1915, vol. II, 1916), says merely that it occupies the ground between algebra and the higher arithmetic, meaning by the latter, as he later explains, what is now called the *Theory of Numbers.*

A current American dictionary (Funk and Wagnalls New Standard, 1943) defines "combinatoric"—a convenient single word which appears now and then in the present text—as "a department of mathematics treating of the formation, enumeration, and properties of partitions, variations, combinations, and permutations of a finite number of elements under various conditions".

The term "combinatorial analysis" itself, seems best explained by the following quotation from Augustus DeMorgan (*Differential and Integral Calculus,* London, 1842, p. 335): "the *combinatorial analysis* mainly consists in the analysis of complicated developments by means of *a priori* consideration and collection of the different combinations of terms which can enter the coefficients".

No one of these statements is satisfactory in providing a safe and sure guide to what is and what is not combinatorial. The authors of the three textbooks could be properly vague because their texts showed what they meant. The dictionary, in describing the contents of such texts, allows no room for new applications of combinatorial technique (such as appear in the last half of Chapter 6 of the present text in the enumeration of trees, networks, and linear graphs). DeMorgan's statement is admirable but half-hearted; in present language, it recognizes that coefficients of generating functions may be determined by solution of combinatorial problems, but ignores the reverse possibility that combinatorial problems may be solved by determining coefficients of generating functions.

Since the subject seems to have new growing ends, and definition is apt to be restrictive, this lack of conceptual precision may be all for the best. So far as the present book is concerned, anything enumerative is combinatorial; that is, the main emphasis throughout is on finding the *number of ways there are* of doing some well-defined operation. This includes all the traditional topics mentioned in the dictionary definition quoted above; therefore this book is suited to the purpose of presenting an introduction to the subject. It is sufficiently vague to include new material, like that mentioned above, and thus it is suited to the purpose of presenting this introduction in an up-to-date form.

The modern developments of the subject are closely associated with the use of generating functions. As appears even in the first chapter, these must be taken in a form more general than the power series given them by P. S. Laplace, their inventor. Moreover, for their combinatorial uses, they are to be regarded, following E. T. Bell, as tools in the theory of an algebra of sequences, so that despite all appearances they belong to algebra and not to analysis. They serve to compress a great deal of development and allow the presentation of a mass of results in a uniform manner, giving the book more scope than would have been possible otherwise. By their means, that central combinatorial tool, the method of inclusion and exclusion, may be shown to be related to the use of factorial moments (which should be attractive to the statistician). Finally, the presentation in this form fits perfectly with the presentation of probability given by William Feller in this series.

As to the contents, the following remarks may be useful. Chapter 1 is a rapid survey of that part of the theory of permutations and combinations which finds a place in books on elementary algebra, with, however, an emphasis on the relation of these results to generating functions which both illuminates and enlarges them. This leads to the extended treatment of generating functions in Chapter 2, where an important result is the introduction of a set of multivariable polynomials, named after their inventor E. T. Bell, which reappear in later chapters. Chapter 3 contains an extended treatment of the principle of inclusion and exclusion which is indispensable to the enumeration of permutations with restricted position given in Chapters 7 and 8. Chapter 4 considers the enumeration of permutations in cyclic representation and may be regarded as an introduction to the beautiful paper, mentioned there, by my friend Jacques Touchard. Chapter 5 is a rapid survey of the theory of distributions, which MacMahon has made almost his own. Chapter 6 considers together partitions, compositions, and the enumeration of trees and linear graphs; much of the latter was developed especially for this book, and is a continuation of work done with my friends, R. M. Foster and C. E. Shannon. Chapters 7 and 8 have been mentioned above and are devoted to elaboration of work at the start of which I was fortunate to have the collaboration of Irving Kaplansky; the continuance of this work owes much to correspondence with both Touchard and Koichi Yamamoto; the development not otherwise accredited is my own.

Each chapter has an extensive problem section, which is intended to carry on the development of the text, and so extend the scope of the book in a way which would have been impossible otherwise. So far as possible, the problems are put in a form to aid rather than baffle the reader; nevertheless, they assume a certain amount of mathematical maturity, that favorite phrase of textbook preface writers, by which I hope they mean a sufficient interest in the subject to do the work necessary to master it. This aid to the reader is also offered in the sequence in which they are set down, which is intended to carry him forward in a natural way.

As to notation, I have limited myself for the most part to English letters, which has entailed using many of these in several senses. So far as possible, these multiple uses have been separated by a decent interval, and warnings have been inserted where confusion seemed likely. The reader now is warned further by this statement. Equations, theorems, sections, examples, and problems are numbered consecutively in each chapter and are referred to by these numbers in their own chapter; in other chapters, their chapter number is prefixed. Thus equation ($3a$) of Chapter 4 is referred to as such in Chapter 4 but as equation ($4.3a$) in Chapter 6. Numbers in bold type following proper names indicate items in the list of references of that chapter. The range of tables is often indicated in the abbreviated fashion currently common: $n = 0(1)10$ means that $n$ has the values $0, 1, 2, \cdots, 10$.

I have made a point of carrying the development to where the actual numbers in question can be obtained as simply as possible; some otherwise barren wastes of algebra are justified by this consideration. In some cases, these numbers are so engaging as to invite consideration of their arithmetical structure (the congruences they satisfy), which I have included, quoting, but not proving, the required results from number theory.

In my first work on this book, which goes back 15 years, I enjoyed an enthusiastic correspondence with Mr. H. W. Becker. In the intensive work of the past 2 years, which would have been impossible without the encouragement of Brockway McMillan, I have had the advantage of presenting the material in two seminars at the Bell Telephone Laboratories. E. N. Gilbert, who arranged one of these, also has given me the benefit of a careful reading of all of the text (in its several versions) and many of the problems, and has also supplied a number of problems. Many improvements have also followed the readings of Jacques Touchard and S. O. Rice. Finally, thanks are due the group of typists under the direction of Miss Ruth Zollo for patient and expert work, and to J. Mysak for the drawings.

JOHN RIORDAN

*Bell Telephone Laboratories*
*New York, N. Y.*
*February, 1958*

# Contents

# Errata

| PAGE | LINE | CORRECTION |
|---|---|---|
| 32 | 14 | $\exp(m(t))$ for $m(t)$ |
| 36 | 7b | (45) for (44) |
| 37 | 13 | $t^2/2!$ for $t^2$ |
| 54 | 8b | Remove half-parenthesis before $S_{k+1}$ |
| 72 | 5b | p. 241 for p. 205 |
| 78 | 16 | **6** for **7** |
| 85 | 7 | $d_{n-r+1}{}^{(t,r)}$ for $d_{n-r+1}(t)$ |
| 151 | 2 | $r(x)$ for $r^2(x)$ |
| 153 | 20 | $[(1-t)(1-t^2)\ldots(1-t^k)]^2$ for $[(1-t)(1-t^2)\ldots(1-t^k)]^k$ |
| 183 | 15 | *Ist. Veneto* for *Inst. Veneto* |
| 217 | 7b | (52) for (51) |
| 218 | 14 | (52) for (51) |
| 218 | 18 | (52) for (51) |
| 222 | 8 | (No. 4, 1950) for (1949) |
| 227 | 20 | Add $(r-1)^n \equiv (r-1)_n$ to $g_n = [(r-1)-2]^n$ |

Notation: line 7b (e.g.) is the seventh line from below.

CHAPTER 1

# Permutations and Combinations

## 1. INTRODUCTION

This chapter summarizes the simplest and most widely used material of the theory of combinations. Because it is so familiar, having been set forth for a generation in textbooks on elementary algebra, it is given here with a minimum of explanation and exemplification. The emphasis is on methods of reasoning which can be employed later and on the introduction of necessary concepts and working tools. Among the concepts is the generating function, the introduction of which leads to consideration of both permutations and combinations in great generality, a fact which seems insufficiently known.

Most of the proofs employ in one way or another either or both of the following rules.

*Rule of Sum:* If object $A$ may be chosen in $m$ ways, and $B$ in $n$ other ways, "either $A$ or $B$" may be chosen in $m + n$ ways.

*Rule of Product:* If object $A$ may be chosen in $m$ ways, and thereafter $B$ in $n$ ways, both "$A$ and $B$" may be chosen *in this order* in $mn$ ways.

These rules are in the nature of definitions (or tautologies) and need to be understood rather than proved. Notice that, in the first, the choices of $A$ and $B$ are mutually exclusive; that is, it is impossible to choose both (in the same way). The rule of product is used most often in cases where the order of choice is immaterial, that is, where the choices are independent, but the possibility of dependence should not be ignored.

The basic definitions of permutations and combinations are as follows:

*Definition. An r-permutation of n things is an ordered selection or arrangement of r of them.*

*Definition. An r-combination of n things is a selection of r of them without regard to order.*

A few points about these should be noted. First, in either case, nothing is said of the features of the $n$ things; they may be all of one kind, some of one kind, others of other kinds, or all unlike. Though in the simpler parts of the theory, they are supposed all unlike, the general case is that of $k$ kinds, with $n_j$ things of the $j$th kind and $n = n_1 + n_2 + \cdots + n_k$. The set of numbers $(n_1, n_2, \cdots, n_k)$ is called the *specification* of the things. Next, in the definition of permutations, the meaning of *ordered* is that two selections are regarded as different if the order of selection is different even when the same things are selected; the $r$-permutations may be regarded as made in two steps, first the selection of all possible sets of $r$ (the $r$-combinations), then the ordering of each of these in all possible ways. For example, the 2-permutations of 3 distinct things labeled 1, 2, 3 are 12, 13, 23; 21, 31, 32; the first three of these are the 2-combinations of these things.

In the older literature, the $r$-permutations of $n$ things are called *variations*, $r$ at a time, for $r < n$, the term permutation being reserved for the ordering operation on all $n$ things, as is natural in the theory of groups, as noticed in Chapter 4. This usage is followed here by taking the unqualified term, permutation, as always meaning $n$-permutation.

## 2. r-PERMUTATIONS

The rules and definition will now be applied to obtain the simplest and most useful enumerations.

### 2.1. Distinct Things

The first of the members of an $r$-permutation of $n$ distinct things may be chosen in $n$ ways, since the $n$ are distinct. This done, the second may be chosen in $n - 1$ ways, and so on until the $r$th is chosen in $n - r + 1$ ways. By repeated application of the rule of product, the number required is

$$P(n, r) = n(n-1)\cdots(n-r+1), \qquad n \geq r \tag{1}$$

If $r = n$, this becomes

$$P(n, n) = n(n-1)\cdots 1 = n! \tag{2}$$

that is, the product of all integers from 1 to $n$, which is called $n$-factorial, is written $n!$ as above. $P(n, 0)$, which has no combinatorial meaning, is taken by convention as unity.

Using (2), (1) may be rewritten

$$P(n, r) = \frac{n!}{(n-r)!} = \frac{P(n, n)}{P(n-r, n-r)}$$

and is also given the abbreviation

$$P(n, r) = (n)_r$$

The last is called the falling $r$-factorial of $n$; since the same notation is also used in the theory of the hypergeometric function to indicate $n(n+1)\cdots(n+r-1)$, the word "falling" in the statement is necessary to avoid ambiguity.

The relation $P(n,n) = P(n, r)P(n-r, n-r)$ appearing above also follows from the rule of product.

It is also interesting to note the following recurrence relation

$$\begin{aligned} P(n, r) &= P(n-1, r) + rP(n-1, r-1) \\ &= (n-1)_r + r(n-1)_{r-1} \end{aligned} \tag{3}$$

This follows by simple algebra from (1), writing $n = n - r + r$; it also follows by classifying the $r$-permutations as to whether they do not or do contain a given thing. If they do not, the number is $P(n-1, r)$; if they do, there are $r$ positions in which the given thing may appear and $P(n-1, r-1)$ permutations of the $n-1$ other things.

*Example 1.* The 12 2-permutations of 4 distinct objects ($n = 4, r = 2$) labeled 1, 2, 3, and 4 are:

12, 13, 14, 23, 24, 34
21, 31, 41, 32, 42, 43

**2.2. The Number of Permutations of $n$ Things, $p$ of Which Are of One Kind, $q$ of Another, etc.**

Here, permutations $n$ at a time only are considered; that is, the term "permutation" is used in the unqualified sense. Let $x$ be the number required. Suppose the $p$ like things are replaced by $p$ new things, distinct from each other and from all other kinds of things being permuted. These may be permuted in $p!$ ways (by equation (2)); hence the number of permutations of the new set of objects is $xp!$. The same goes for every other set of like things and, since finally all things become unlike and the number of permutations of $n$ unlike things is $n!$,

$$xp!\,q!\cdots = n!$$

or

$$x = \frac{n!}{p!\,q!\cdots}, \qquad p + q + \cdots = n \tag{4}$$

The right of (4) is a multinomial coefficient, that is, a coefficient in the expansion of $(a + b + \cdots)^n$. It is also the number of arrangements of $n$ unlike things into unlike cells or boxes, $p$ in the first, $q$ in the second, and so on, without regard to order in any cell, as will be shown in Chapter 5.

*Example 2.* The number of arrangements on one shelf of $n$ different books, for each of which there are $m$ copies, is $(nm)!/(m!)^n$.

The more general problem of finding the number of $r$-permutations of things of general specification ($p$ of one kind, $q$ of another, etc.) will be treated in a later section of this chapter by the method of generating functions.

### 2.3. $r$-Permutations with Unrestricted Repetition

Here there are $n$ kinds of things, and an unlimited number of each kind or, what is the same thing, any object after being chosen is replaced in the stock. Hence, each place in an $r$-permutation may be filled in $n$ different ways, because the stock of things is unaltered by any choice. By the rule of product, the number of permutations in question, that is, the $r$-permutations with repetition of $n$ things, is,

$$U(n, r) = n^r \tag{5}$$

*Example 3.* The number of ways an $r$-hole tape (as used in teletype and computing machinery) can be punched is $2^r$. Here there are two "objects", clear (unpunched) and punched tape, for each of the $r$ positions of any line on the tape.

In the language of statistics, $U(n, r)$ is for sampling with replacement, as contrasted with $P(n, r)$ which is for sampling without replacement.

## 3. COMBINATIONS

### 3.1. $r$-Combinations of $n$ Distinct Things

The simplest derivation is to notice that each combination of $r$ distinct things may be ordered in $r!$ ways, and so ordered is an $r$-permutation; hence, if $C(n, r)$ is the number in question

$$r!\, C(n, r) = P(n, r) = n(n-1)\cdots(n-r+1), \qquad n \geq r$$

and

$$C(n, r) = \frac{n(n-1)\cdots(n-r+1)}{r!} = \frac{n!}{r!\,(n-r)!} = \binom{n}{r} \tag{6}$$

where the last symbol is that usual for binomial coefficients, that is, the coefficients in the expansion of $(a+b)^n$. ($C_r^n$, $C_n^r$, ${}_nC_r$, and $(n, r)$ are alternative notations; the first two are difficult in print and the superscripts are liable to confusion with exponents, the first subscript of the third is sometimes hard to associate with the $C$ which it modifies, the $C$ is strictly unnecessary since the numbers are not necessarily associated with combinations, but all have the sanction of some usage in and out of print.)

$C(n, 0)$, the number of combinations of $n$ things zero at a time, has no combinatorial meaning, but is given by (6) to be unity, which is the usual convention. The values

$$C(n, r) = 0, \qquad r < 0 \text{ and } r > n$$

are also in agreement with (6). $C(-n, r)$ has no combinatorial meaning, but it may be noticed that, from (6)

$$\binom{-n}{r} = (-1)^r \binom{n + r - 1}{r}$$

and

$$\binom{-n}{-r} = (-1)^{n+r} \binom{r - 1}{n - 1}$$

*Example 4.* The 6 combinations of 4 distinct objects taken two at a time ($n = 4, r = 2$), labeling the objects 1, 2, 3, and 4, are:

12, 13, 14, 23, 24, 34

Another derivation of (6) is by recurrence and the rule of sum. The combinations may be divided into those which include a given thing, say the first, and those which do not. The number of those of the first kind is $C(n - 1, r - 1)$, since fixing one element of a combination reduces both $n$ and $r$ by one; the number of those of the second kind, for a similar reason, is $C(n - 1, r)$. Hence,

$$C(n, r) = C(n - 1, r - 1) + C(n - 1, r), \qquad n \geq r \tag{7}$$

Using mathematical induction, it may be shown that (6) is the only solution of (7) which satisfies the boundary conditions $C(n, 0) = C(1, 1) = 1$, $C(1, 2) = C(1, 3) = \cdots = 0$.

Equation (7) is an important formula because it is the basic recurrence for binomial coefficients. Notice that by iteration it leads to

$$\begin{aligned} C(n,r) &= C(n-1,r-1) + C(n-2,r-1) + \cdots + C(r-1,r-1) \qquad (8) \\ &= C(n - 1, r) + C(n - 2, r - 1) + \cdots + C(n - 1 - r, 0) \end{aligned}$$

The first of these classifies the $r$-combinations of $n$ numbered things according to the size of the smallest element, that is, $C(n - k, r - 1)$ is the number of $r$-combinations in which the smallest element is $k$. The similar combinatorial proof of the second does not lend itself to simple statement.

For numerical concreteness, a short table of the numbers $C(n, r)$, a version of the Pascal triangle, is given below. Note how easily one row may be filled in from the next above by (7). Note also the symmetry, that is $C(n, r) = C(n, n - r)$; this follows at once from (6), and is also evident from the fact that a selection of $r$ determines the $n$–$r$ elements not selected.

NUMBERS $C(n, r)$ (BINOMIAL COEFFICIENTS)

| $n\backslash r$ | 0 | 1 | 2 | 3 | 4 | 5 |
|---|---|---|---|---|---|---|
| 0 | 1 | | | | | |
| 1 | 1 | 1 | | | | |
| 2 | 1 | 2 | 1 | | | |
| 3 | 1 | 3 | 3 | 1 | | |
| 4 | 1 | 4 | 6 | 4 | 1 | |
| 5 | 1 | 5 | 10 | 10 | 5 | 1 |

### 3.2. Combinations with Repetition

The number sought is that of $r$-combinations of $n$ distinct things, each of which may appear indefinitely often, that is, 0 to $r$ times. This is a function of $n$ and $r$, say $f(n, r)$.

The most natural method here seems to be that of recurrence and the rule of sum. Suppose the things are numbered 1 to $n$; then the combinations either contain 1 or they do not. If they do, they may contain it once, twice, and so on, up to $r$ times, but, in any event, if one of the appearances of element 1 is crossed out, $f(n, r - 1)$ possible combinations are left. If they do not, there are $f(n - 1, r)$ possible combinations. Hence,

$$f(n, r) = f(n, r - 1) + f(n - 1, r) \tag{9}$$

If $r = 1$, no repetition of elements is possible and $f(n, 1) = n$. (Note that this determines $f(n, 0) = f(n, 1) - f(n - 1, 1)$ as 1, a natural convention.) If $n = 1$, only one combination is possible whatever $r$, so $f(1, r) = 1$. Now

$$\begin{aligned} f(n, 2) &= f(n, 1) + f(n - 1, 2) \\ &= f(n, 1) + f(n - 1, 1) + f(n - 2, 2) \end{aligned}$$

and if this is repeated until the appearance of $f(1, 2)$, which is unity,

$$\begin{aligned} f(n, 2) &= n + (n - 1) + (n - 2) + \cdots + 1 \\ &= (n + 1)n/2 = \binom{n+1}{2} \end{aligned}$$

by the formula for the sum of an arithmetical series, or by the first of equations (8).

Similarly

$$\begin{aligned} f(n, 3) &= f(n, 2) + f(n-1, 2) + \cdots + f(1, 3) \\ &= \binom{n+1}{2} + \binom{n}{2} + \cdots + 1 \\ &= \binom{n+2}{3} \end{aligned}$$

again by the first of equations (8).

The form of the general answer, that is, the number of $r$-combinations with repetition of $n$ distinct things, now is clearly

$$f(n, r) = \binom{n+r-1}{r} \tag{10}$$

and it may be verified without difficulty that this satisfies (9) and its boundary conditions: $f(n, 1) = n, f(1, r) = 1$.

*Example 5.* The number of combinations with repetition, 2 at a time, of 4 things numbered 1 to 4, by equation (10), is 10; divided into two parts as in equation (9), these combinations are

11, 12, 13, 14
22, 23, 24, 33, 34, 44

The result in (10) is so simple as to invite proofs of equal simplicity. The best of these, which may go back to Euler, is as follows. Consider any one of the $r$-combinations with repetition of $n$ numbered things, say, $c_1c_2 \cdots c_r$ in rising order (with like elements taken to be rising), where, of course, because of unlimited repetition, any number of consecutive $c$'s may be alike. From this, form a set $d_1d_2 \cdots d_r$ by the rules: $d_1 = c_1 + 0$, $d_2 = c_2 + 1, d_i = c_i + i - 1, \cdots, d_r = c_r + r - 1$; hence, whatever the $c$'s, the $d$'s are unlike. It is clear that the sets of $c$'s and $d$'s are equinumerous; that is, each distinct $r$-combination produces a distinct set of $d$'s, and vice versa. The number of sets of $d$'s is the number of $r$-combinations (without repetition) of elements numbered 1 to $n + r - 1$, which is $C(n + r - 1, r)$, in agreement with (10); for example, the $d$'s corresponding to the $r$-combinations of Example 5 are (in the same order)

12, 13, 14, 15
23, 24, 25, 34, 35, 45

## 4. GENERATING FUNCTION FOR COMBINATIONS

The enumerations given above may be unified and generalized by a relatively simple mathematical device, the generating function.

For illustration, consider three objects labeled $x_1$, $x_2$, and $x_3$. Form the algebraic product

$$(1 + x_1t)(1 + x_2t)(1 + x_3t)$$

Multiplied out and arranged in powers of $t$, this is

$$1 + (x_1 + x_2 + x_3)t + (x_1x_2 + x_1x_3 + x_2x_3)t^2 + x_1x_2x_3t^3$$

or, in the notation of symmetric functions,

$$1 + a_1t + a_2t^2 + a_3t^3$$

where $a_1$, $a_2$, and $a_3$ are the elementary symmetric functions of the three variables $x_1$, $x_2$, and $x_3$. These symmetric functions are identified by the equation ahead, and it will be noticed that $a_r$, $r = 1, 2, 3$, contains one term for each combination of the three things taken $r$ at a time. Hence, the number of such combinations is obtained by setting each $x_i$ to unity, that is,

$$(1 + t)^3 = \sum_{r=0}^{3} a_r(1, 1, 1)t^r$$

In the case of $n$ distinct things labeled $x_1$ to $x_n$, it is clear that

$$\begin{aligned}(1 + x_1t)(1 + x_2t)\cdots(1 + x_nt) = 1 &+ a_1(x_1, x_2, \cdots, x_n)t + \cdots \\ &+ a_r(x_1, x_2, \cdots, x_n)t^r + \cdots \\ &+ a_n(x_1, x_2, \cdots, x_n)t^n\end{aligned}$$

and

$$(1 + t)^n = \sum_{r=0}^{n} a_r(1, 1, \cdots, 1)t^r = \sum_{r=0}^{n} C(n, r)t^r \tag{11}$$

a result foreshadowed in the remark following (6), which identifies the numbers $C(n, r)$ as binomial coefficients.

The expression $(1 + t)^n$ is called the enumerating generating function or, for brevity, simply the enumerator, of combinations of $n$ distinct things.

The effectiveness of the enumerator in dealing with the numbers it enumerates is indicated in the examples following.

*Example 6.* In equation (11), put $t = 1$; then

$$2^n = \sum_0^n C(n, r) = \sum_0^n \binom{n}{r}$$

that is, the total number of combinations of $n$ distinct things, any number at a time, is $2^n$, which is otherwise evident by noticing that in this total each element either does or does not appear. With $t = -1$, equation (11) becomes

$$\begin{aligned}0 &= \sum_0^n (-1)^r \binom{n}{r} \\ &= 1 - n + \binom{n}{2} - \binom{n}{3} + \cdots + (-1)^n \binom{n}{n}\end{aligned}$$

Adding and subtracting these equations leads to

$$\sum_{r=0} \binom{n}{2r} = \sum_{r=0} \binom{n}{2r+1} = 2^{n-1}$$

*Example 7.* Write $n = n - m + m$. Since

$$(1 + t)^n = (1 + t)^{n-m}(1 + t)^m$$

it follows by equating coefficients of $t^r$ that

$$C(n, r) = C(n - m, 0)C(m, r) + C(n - m, 1)C(m, r - 1) + \cdots + C(n - m, r)C(m, 0)$$

or

$$\binom{n}{r} = \sum_{k=0} \binom{n-m}{k} \binom{m}{r-k}$$

The corresponding relation for falling factorials $(n)_r = n(n - 1) \cdots (n - r + 1)$ is

$$\begin{aligned}(n)_r = (m)_r + \binom{r}{1}(n - m)_1(m)_{r-1} + \binom{r}{2}(n - m)_2(m)_{r-2} + \cdots \\ + \binom{r}{j}(n - m)_j(m)_{r-j} + \cdots \\ + (n - m)_r\end{aligned}$$

This is often called Vandermonde's theorem.

Another form is

$$(n)_r = \sum_{j=0} \binom{r}{j}(n + m)_j(-m)_{r-j}$$

which implies the relation in rising factorials:

$$n^{(r)} = n(n + 1) \cdots (n + r - 1) = (-1)^r(-n)_r$$

namely,

$$n^{(r)} = \sum_{j=0} \binom{r}{j}(n - m)^{(j)}m^{(r-j)}$$

The result in (11) is only a beginning. What are the generating functions and enumerators when the elements to be combined are not distinct?

In the expression $(1 + x_1t)(1 + x_2t) \cdots (1 + x_nt)$, each factor of the product is a binomial (2-termed expression) which indicates in the terms 1 and $x_kt$ the fact that element $x_k$ may not or may appear in any combination. The product generates combinations because the coefficient of $t^r$ is obtained by picking unity terms from *n-r* factors and terms like $x_kt$ from the $r$ remaining factors in all possible ways; these are the $r$-combinations by definition. The factors are limited to two terms because no object may appear more than once in any combination.

Hence, if the combinations may include object $x_k$ 0, 1, 2, $\cdots j$ times, the generating function is altered by writing

$$1 + x_kt + x_k^2t^2 + \cdots + x_k^jt^j$$

in place of $1 + x_k t$. Moreover, the factors may be tailored to any specifications quite independently. Thus, if $x_k$ is always to appear an even number of times but not more than $j$ times, the factor (with $j = 2i + 1$) is

$$1 + x_k^2 t^2 + x_k^4 t^4 + \cdots + x_k^{2i} t^{2i}$$

Hence the generating function of any problem describes not only the kinds of objects but also the kinds of combinations in question. A factor $x_k^i$ in any term of a coefficient of a power of $t$ in the generating function indicates that object $x_k$ appears $i$ times in the corresponding combination.

A general formula for all possibilities could be written down, but the notation would be cumbersome and the formula probably less illuminating than the examples which follow.

*Example 8.* For combinations with unlimited repetition of objects of $n$ kinds and no restriction on the number of times any object may appear, the enumerating generating function is

$$(1 + t + t^2 + \cdots)^n$$

This is the same as

$$\begin{aligned}(1 - t)^{-n} &= \sum_0^\infty \binom{-n}{r} (-t)^r \\ &= \sum_0^\infty \frac{-n(-n-1)\cdots(-n-r+1)}{r!} (-t)^r \\ &= \sum_0^\infty \binom{n+r-1}{r} t^r\end{aligned}$$

which confirms the result in (10).

*Example 9.* For combinations as in Example 8 and the further condition that at least one object of each kind must appear, the enumerating generating function is

$$(t + t^2 + \cdots)^n$$

and

$$\begin{aligned}(t + t^2 + \cdots)^n &= t^n(1 - t)^{-n} \\ &= t^n \sum_0^\infty \binom{n+r-1}{r} t^r \\ &= \sum_{r=n}^\infty \binom{r-1}{r-n} t^r = \sum_{r=n} \binom{r-1}{n-1} t^r\end{aligned}$$

that is, the number of combinations in question is 0 for $r < n$ and $C(r - 1, r - n)$ for $r \geq n$. For instance, for $n = 3$ and elements $a$, $b$, $c$, there is one 3-combination $abc$ and six $= C(4, 2)$ 5-combinations, namely, *aaabc*, *abbbc*, *abccc*, *aabbc*, *aabcc*, *abbcc*.

*Example 10.* For combinations as in Example 8, but with each object appearing an even number of times, the enumerator is

$$(1 + t^2 + t^4 + \cdots)^n$$

which is the same as

$$(1 - t^2)^{-n} = \sum_0^\infty \binom{n + r - 1}{r} t^{2r}$$

Thus the number of $r$-combinations for $r$ odd is zero, and the $2r$-combinations are equinumerous with the $r$-combinations of Example 8, as is immediately evident. Note that

$$(1 - t^2)^{-n} = (1 - t)^{-n}(1 + t)^{-n}$$

so that the sum

$$\binom{n + r - 1}{r} - \binom{n + r - 2}{r - 1} n + \binom{n + r - 3}{r - 2}\binom{n + 1}{2} \cdots$$
$$+ (-1)^k \binom{n + r - k - 1}{r - k}\binom{n + k - 1}{k} + \cdots + (-1)^r \binom{n + r - 1}{r}$$

is zero for $r$ odd and equal to $\binom{n + s - 1}{s}$ for $r = 2s$, that is,

$$\binom{n + s - 1}{s} = \sum_{k=0}^{2s} \binom{n + 2s - k - 1}{2s - k}\binom{n + k - 1}{k} (-1)^k$$

Obviously, examples like these could be multiplied indefinitely. The essential thing to notice is that, in forming a combination, the objects are chosen independently and the generating function takes advantage of this independence by a rule of multiplication. In effect, each factor of a product is a generating function for the objects of a given kind. These generating functions appear in product just as they do for the sum of independent random variables in probability theory.

## 5. GENERATING FUNCTIONS FOR PERMUTATIONS

Since $x_1x_2$ and $x_2x_1$ are indistinguishable in an algebraic commutative process, it is impossible to give a generating function which will exhibit permutations as those above have exhibited combinations. Nevertheless, the enumerators are easy to find.

For $n$ unlike things, it follows at once from (1) that

$$(1 + t)^n = \sum_0^n P(n, r)t^r/r! \tag{12}$$

that is, $(1 + t)^n$ on expansion gives $P(n, r)$ as coefficient of $t^r/r!$.

This supplies the hint for generalization. If an element may appear $0, 1, 2, \cdots, k$ times, or if there are $k$ elements of a given kind, a factor $1 + t$ on the left of (12) is replaced by

$$1 + t + \frac{t^2}{2!} + \cdots + \frac{t^k}{k!}$$

This is because the number of permutations of $n$ things, $p$ of which are of one kind, $q$ of another, and so on, is given by (4) as

$$\frac{n!}{p!\,q!\cdots}$$

which is the coefficient of $t^n/n!$ in the product

$$\frac{t^p}{p!}\frac{t^q}{q!}\cdots, \qquad p+q+\cdots=n$$

which corresponds to the prescription that the letter indicating things of the first kind appear exactly $p$ times, the letter for things of the second kind appear exactly $q$ times, and so on.

Hence, if the permutations are prescribed by the conditions that the $k$th of $n$ elements is to appear $\lambda_0(k), \lambda_1(k), \cdots$ times, $k = 1$ to $n$, the number of $r$-permutations is the coefficient of $t^r/r!$ in the product

$$\prod_{k=1}^{n}\left(\frac{t^{\lambda_0(k)}}{\lambda_0(k)!}+\frac{t^{\lambda_1(k)}}{\lambda_1(k)!}+\cdots\right)$$

The (enumerating) generating functions appearing here may be called exponential generating functions, as suggested by

$$e^{at}=\sum_0^\infty a^r\frac{t^r}{r!}$$

These remarks will become clearer in the examples which follow.

*Example 11.* For $r$-permutations of $n$ different objects with unlimited repetition (no restriction on the number of times any object may appear), the enumerator is

$$\left(1+t+\frac{t^2}{2!}+\cdots\right)^n$$

But

$$\left(1+t+\frac{t^2}{2!}+\cdots\right)^n = e^{nt}$$

$$= \sum_0^\infty n^r\frac{t^r}{r!}$$

Hence, the number of $r$-permutations is $n^r$, in agreement with (5).

*Example 12.* For permutations as in Example 11 and the additional condition that each object must appear at least once, the enumerator is

$$\left(t + \frac{t^2}{2!} + \cdots\right)^n = (e^t - 1)^n$$

$$= \sum_{j=0}^{n} \binom{n}{j} (-1)^j e^{(n-j)t}$$

$$= \sum_{r=0}^{\infty} \frac{t^r}{r!} \sum_{0}^{n} \binom{n}{j} (-1)^j (n-j)^r$$

$$= \sum_{r=0}^{\infty} \frac{t^r}{r!} \Delta^n 0^r$$

The last result is in the notation of the calculus of finite differences. If $E$ is a shift operator such that $Eu(n) = u(n+1)$ and $\Delta$ is the difference operator defined by $\Delta u(n) = u(n+1) - u(n) = (E-1)u(n)$, the result may be obtained as follows:

$$\Sigma \binom{n}{j} (-1)^j (n-j)^r = \Sigma \binom{n}{j} (-1)^j E^{n-j} 0^r$$

$$= (E-1)^n 0^r = \Delta^n 0^r$$

For the sake of being concrete, it may be noted that $\Delta 0^r = 1^r$, $\Delta^2 0^r = 2^r - 2$, $\Delta^3 0^r = 3^r - 3 \cdot 2^r + 3$. These numbers appear again in later chapters.

*Example 13.* For $r$-permutations of elements with the specification of Section 2.2, that is, $p$ of one kind, $q$ of a second, and so on, the enumerating generating function is

$$(1 + t + \cdots + t^p/p!)(1 + t + \cdots + t^q/q!) \cdots$$

It is easy to verify (compare the remark on page 12) that the coefficient of $t^n/n!$ is $n!/p!\,q! \cdots$, in agreement with (4). The number of $r$-permutations is the coefficient of $t^r/r!$; hence the generating function above is the result promised at the end of Section 2.2.

Finally it may be noted that the permutations with unlimited repetition of Examples 11 and 12 may be related to problems of distribution which will be treated in Chapter 5. Take for illustration the permutations three at a time and with repetition of two kinds of things, say $a$ and $b$, which are eight in number, namely

$$aaa \quad aab \quad aba \quad baa \quad abb \quad bab \quad bba \quad bbb$$

These may be put into 1–1 correspondence with the distribution of 3 different objects into 2 cells, as indicated by

$$abc|— \quad ab|c \quad ac|b \quad bc|a \quad a|bc \quad b|ac \quad c|ab \quad —|abc$$

the vertical line separating the cells.

This suggests what will be shown in Chapter 5, namely: the permutations with repetition of objects of $n$ kinds $r$ at a time are equinumerous with the distributions of $r$ different objects into $n$ different cells. The restriction of Example 12 that each permutation contain at least one object of each kind corresponds to the restriction that no cell may be empty.

## REFERENCES

1. S. Barnard and J. M. Child, *Higher Algebra*, New York, 1936.
2. G. Chrystal, *Algebra* (Part II), London, 1900.
3. E. Netto, *Lehrbuch der Combinatorik*, Leipzig, 1901.
4. W. A. Whitworth, *Choice and Chance*, London, 1901.

## PROBLEMS

1. (*a*) Show that $p$ plus signs and $q$ minus signs may be placed in a row so that no two minus signs are together in $\binom{p+1}{q}$ ways.

(*b*) Show that $n$ signs, each of which may be plus or minus, may be placed in a row with no two minus signs together in $f(n)$ ways, where $f(0) = 1$, $f(1) = 2$ and

$$f(n) = f(n-1) + f(n-2), \qquad n > 1$$

(*c*) Comparison of (*a*) and (*b*) requires that

$$f(n) = \sum_{q=0}^{m} \binom{n-q+1}{q}, \qquad m = [(n+1)/2]$$

with $[x]$ indicating the largest integer not greater than $x$. Show that

$$g(n) = \sum_{q=0}^{m} \binom{n-q+1}{q}$$
$$= g(n-1) + g(n-2)$$

and $g(0) = 1$, $g(1) = 2$, so that $g(n) = f(n)$. The numbers $f(n)$ are Fibonacci numbers.

2. (*a*) Show that

$$n\binom{n}{r} = (r+1)\binom{n}{r+1} + r\binom{n}{r}$$

$$\binom{n}{2}\binom{n}{r} = \binom{r+2}{2}\binom{n}{r+2} + 2\binom{r+1}{2}\binom{n}{r+1} + \binom{r}{2}\binom{n}{r}$$

$$\binom{n}{s}\binom{n}{r} = \sum_{k=0}^{q}\binom{s}{k}\binom{r+s-k}{r-k}\binom{n}{r+s-k}, \qquad q = \min(r, s)$$

(*b*) Show similarly that

$$n\binom{n}{r} = r\binom{n+1}{r+1} + \binom{n}{r+1}$$

$$\binom{n}{2}\binom{n}{r} = \binom{r}{2}\binom{n+2}{r+2} + 2r\binom{n+1}{r+2} + \binom{n}{r+2}$$

$$\binom{n}{s}\binom{n}{r} = \sum_{k=0}\binom{s}{k}\binom{r}{s-k}\binom{n+s-k}{r+s}$$

3. From the generating function $(1+t)^n = \sum C(n, r)t^r$, or otherwise, show that

$$\sum_{r=1}^{n} \frac{(-1)^{r+1}}{r+1} C(n, r) = \frac{n}{n+1}$$

$$\sum_{r=0}^{n} \frac{1}{r+1} C(n, r) = \frac{2^{n+1}-1}{n+1}$$

4. Derive the following binomial coefficient identities:

$$\sum_{k=0}^{m} (-1)^k \binom{n-k}{n-m}\binom{n}{k} = 0, \qquad 0 < m \leq n$$

$$\sum_{k=0}^{m} \binom{n-k}{n-m}\binom{n}{k} = 2^m\binom{n}{m}$$

$$\sum_{k=0}^{n-m} (-1)^k \binom{n-k}{m}\binom{n}{k} = 0, \qquad 0 \leq m < n$$

$$\sum_{k=0}^{m} \binom{n}{k}\binom{n}{m-k} = \binom{2n}{m}, \qquad m \leq n; \qquad \sum_{k=0}^{n} \binom{n}{k}^2 = \binom{2n}{n}$$

5. Show that the number of $r$-combinations of $n$ objects of specification $p1^q(p+q=n)$, that is, with $p$ of one kind and one each of $q$ other kinds, is

$$C(p1^q; r) = \binom{q}{r} + \binom{q}{r-1} + \cdots + \binom{q}{r-p}$$

From this, show that the enumerator is

$$(1 + t + t^2 + \cdots + t^p)(1+t)^q$$

and determine the recurrence

$$C(p1^q, r) - C(p1^q, r-1) = \binom{q}{r} - \binom{q}{r-p-1}$$

6. Show that the number of $r$-combinations of $n$ objects of specification $2^m1^{n-2m}$ is

$$C(2^m1^{n-2m}, r) = \sum_k \binom{m}{k}\binom{n-m-k}{r-2k}$$

Derive the recurrence

$$C(2^m1^{n-2m}, r) = C(2^{m-1}1^{n+1-2m}, r) + C(2^{m-1}1^{n-2m}, r-2)$$

7. Show that the number of $r$-combinations of objects of specification $s^m$ ($m$ kinds of objects, $s$ of each kind), sometimes called "regular" combinations, is

$$C(s^m, r) = \sum_k (-1)^k \binom{m}{k} \binom{m + r - k(s+1) - 1}{r - k(s+1)}$$

8. Write $C(n_1 n_2 \cdots n_m, r)$ for the number of $r$-combinations of objects of specification $(n_1 n_2 \cdots n_m)$; derive the recurrence

$$C(n_1 n_2 \cdots n_m, r) = C(n_2 \cdots n_m, r) + C(n_2 \cdots n_m, r-1) + \cdots + C(n_2 \cdots n_m, r - n_1)$$

9. For permutations with unlimited repetition, derive the result of Example 12 by classifying the permutations of Example 11, $n^r$ in total, according to the number of distinct objects they contain.

10. From the generating function of Example 12, or otherwise, show that

$$\Delta^n 0^{r+1} = n\,\Delta^n 0^r + n\,\Delta^{n-1} 0^r$$

11. For permutations with unlimited repetition and each object restricted to appearing zero or an even number of times, show that the number of $r$-permutations may be written $\delta^n 0^r$ with $\delta u_n = (u_{n+1} + u_{n-1})/2$. Derive the recurrence

$$\delta^n 0^{r+2} = n^2\,\delta^n 0^r - n(n-1)\,\delta^{n-2} 0^r$$

12. Write $P_n$ for the total number of permutations of $n$ distinct objects; that is, $P_n = \Sigma\,(n)_r$.

($a$) Derive the recurrence

$$P_n = nP_{n-1} + 1$$

and verify the table

| $n$ | 0 | 1 | 2 | 3 | 4 | 5 | 6 | 7 | 8 |
|---|---|---|---|---|---|---|---|---|---|
| $P_n$ | 1 | 2 | 5 | 16 | 65 | 326 | 1957 | 13700 | 109601 |

($b$) Show that $P_n = (E+1)^n 0! = n! \sum_0^n 1/k! = n!e - \dfrac{1}{n+1} - \dfrac{1}{(n+2)_2} - \cdots$ and, hence, that $P_n$ is the nearest integer to $n!e$.

($c$) Show that

$$\sum_{n=0}^{\infty} P_n t^n/n! = e^t/(1-t)$$

13. For objects of type $(2^m)$, that is, of $m$ kinds and two of each kind, the enumerator for $r$-permutations is

$$\left(1 + t + \frac{t^2}{2}\right)^m = \sum_{r=0}^{2m} q_{m,r} t^r/r!$$

where $q_{m,r}$ is the number of $r$-permutations.

(*a*) Derive the following recurrences:

$$q_{m+1,r} = q_{m,r} + rq_{m,r-1} + \binom{r}{2} q_{m,r-2}$$

$$q_{m,r+1} = mq_{m,r} - m\binom{r}{2}q_{m-1,r-2}$$

$$q_{m,r+2} = m(2m-1)q_{m-1,r} - m(m-1)q_{m-2,r}$$

Verify the particular results $q_{m,0} = 1$, $q_{m,1} = m$, $q_{m,2} = m^2$.

(*b*) For the total number of permutations, $q_m = \Sigma\, q_{m,r}$, derive the symbolic expression

$$q_m = (1 + E + E^2/2)^m 0!, \qquad (E^k 0! = k!)$$

and the recurrence relation

$$q_m = m(2m-1)q_{m-1} - m(m-1)q_{m-2} + m + 1$$

Verify that $q_0 = 1$, $q_1 = 3$, $q_2 = 19$, $q_3 = 271$.

14. By classifying the permutations with repetition of $n$ distinct objects, show that

$$r^r(n-r)^{n-r}\binom{n}{r} \leq n^n$$

or

$$\binom{n}{r} \leq n^n/r^r(n-r)^{n-r}$$

15. Terquem's problem. For combinations of $n$ numbered things in natural (rising) order, and with $f(n, r)$ the number of $r$-combinations with odd elements in odd position and even elements in even position, or, what is the same thing, with $f(n, r)$ the number of combinations with an equal number of odd and even elements for $r$ even and with the number of odd elements one greater than the number of even for $r$ odd,

(*a*) Show that $f(n, r)$ has the recurrence

$$f(n, r) = f(n-1, r-1) + f(n-2, r), \qquad f(n, 0) = 1$$

(*b*) $$f(n, r) = \binom{q}{r}, \qquad q = [(n+r)/2]$$

$[(n + r)/2]$ being the greatest integer not greater than $(n + r)/2$.

(*c*) $$f(n) = \Sigma\, f(n, r) = f(n-1) + f(n-2)$$

The numbers $f(n)$ are Fibonacci numbers, as in Problem 1.

16. Write $a_{n,r}$ for the number of $r$-permutations with repetition of $n$ different things, such that no two consecutive things are alike, and $a^*_{n,r}$ for the number of these with a given element first. Show that

$$a_{n,r} = na^*_{n,r}$$
$$a^*_{n,r} = (n-1)a^*_{n,r-1}$$

and hence

$$a_{n,r} = (n-1)a_{n,r-1}, \qquad r > 1$$
$$= n(n-1)^{r-1}, \qquad r > 0$$

With

$$a_n(x) = \sum_{r=1} a_{n,r} x^{r-1}$$

show that

$$a_n(x) = n[1 - x(n-1)]^{-1}$$

17. (*a*) Write $b_{n,r}$ for permutations as in Problem 16 except that no three consecutive things are alike. By use of the corresponding numbers with a given thing first and a given pair of things first, derive the recurrence

$$b_{n,r} = (n-1)(b_{n,r-1} + b_{n,r-2}), \qquad r > 2$$

From this and the initial values $b_{n,1} = n$, $b_{n,2} = n^2$, derive the generating function relation

$$[1 - (n-1)(x + x^2)]b_n(x) = n(1 + x)$$

with

$$b_n(x) = \sum_{r=1} b_{n,r} x^{r-1}$$

(*b*) Show that

$$1 + \frac{n-1}{n} x b_n(x) = \frac{1}{1 - (n-1)(x + x^2)}$$
$$= \sum_{r=0} \frac{\alpha^{r+1} - \beta^{r+1}}{\alpha - \beta} x^r$$

with

$$(1 - \alpha x)(1 - \beta x) = 1 - (n-1)(x + x^2)$$

Hence,

$$\frac{n-1}{n} b_{n,r} = \frac{\alpha^{r+1} - \beta^{r+1}}{\alpha - \beta}$$

18. For permutations as in Problems 16 and 17 and no $m$ consecutive things alike, $m = 2, 3, \cdots$ write $a_{n,r}^{(m)}$ and $a_n^{(m)}(x)$ for the numbers and generating functions. Show that

$$a_{n,r}^{(m)} = (n-1)(a_{n,r-1}^{(m)} + a_{n,r-2}^{(m)} + \cdots + a_{n,r-m+1}^{(m)}), \qquad r > m - 1$$

and, since $a_{n,r}^{(m)} = n^r$, $r < m$,

$$[1 - (n-1)(x + x^2 + \cdots + x^{m-1})]a_n^{(m)}(x) = n(1 + x + \cdots + x^{m-2})$$

CHAPTER 2

# Generating Functions

## 1. INTRODUCTION

It is clear from the discussion in Chapter 1 that a generating function of some form may be an important means of unifying the treatment of combinatorial problems. This could have been predicted from the De Morgan definition of combinatorial analysis cited in the preface, for if the latter is a means of finding coefficients in complicated developments of given functions, then these functions may be regarded as generating the coefficients, and their study is the natural complement to the study of the coefficients. The beginnings of this study are examined in this chapter.

First, the informal discussion of Chapter 1 of the two kinds of generating function, one for combinations, the other for permutations, may now be replaced by the following definitions:

(*a*) The ordinary *generating function* of the sequence $a_0, a_1, \cdots$ is the sum

$$A(t) = a_0 + a_1 t + a_2 t^2 + \cdots + a_n t^n + \cdots \tag{1}$$

(*b*) The *exponential generating function* of the same sequence is the sum

$$E(t) = a_0 + a_1 t + a_2 t^2/2! + \cdots + a_n t^n/n! + \cdots \tag{2}$$

Several remarks on these definitions are necessary. First, as indicated by the subscripts, the sequence $a_0, a_1, \cdots$ is ordered and its elements may or may not be distinct. Second, the variable $t$ of the generating function has not been defined; the natural tendency of the classical analyst, and the pioneers of the subject, to take it as a real or complex number need not be followed. If it is, the question of the existence of the generating function for a given infinite sequence arises at once and, of course, must be answered, that is, the convergence of the sum defining the generating function must be established. It may be noted that, when the sequence is bounded, that is, when $|a_n| \leq M$, $n = 0, 1, 2, \cdots$, where $M$ is some non-infinite number, the sum defining $A(t)$ converges for $|t| < 1$, whereas that defining $E(t)$ converges for all $t$. On the other hand, when the variable $t$ is taken as an abstract mark or indeterminate, the function of which is to keep distinct through its powers the elements of the sequence united in the sum defining the generating function, the latter becomes a tool in an algebra of these sequences. Formal operations on generating functions, such as addition, multiplication, differentiation, and integration with respect to $t$, serve to express definitions and relations in this algebra, in particular by equating coefficients of $t^n$ after completing these operations. From this point of view, convergence of the sum is irrelevant. (A formal and detailed expression of this point of view may be found in references **1**, **2**, and **3** (E. T. Bell).)

The algebra associated with the power series generating function $A(t)$ is known as the Cauchy algebra; that associated with the exponential generating function $E(t)$ is known as the Blissard (or symbolic, or umbral) calculus. $E(t)$ is less familiar than $A(t)$; indeed, some hesitate to call it a generating function at all, reserving the term exclusively for $A(t)$. However, its appearance in Chapter 1 already attests to its combinatorial usefulness; it appears in statistics as a moment generating function, and it has also been used in number theory (Lucas, **7**, Bell, **2**). Moreover, as will be shown, generating functions different from either are inevitable in some parts of the subject.

In an introductory book like the present one, it seems unnecessary to burden the text with the elaboration of definition and explanation required for formal completeness. Instead, formal operations are used freely in an heuristic manner with verifications by independent arguments wherever possible within space limits.

Next, to complete the story of generating functions, it should be noticed first that for a single variable both $A(t)$ and $E(t)$ are instances of the expression

$$G(t) = a_0 f_0(t) + a_1 f_1(t) + \cdots + a_n f_n(t) + \cdots \tag{3}$$

the sole requirement for which is that the functions $f_0(t), f_1(t), \cdots$ be linearly independent (in order to make the expression unique). The corresponding expression for a 2-index sequence $a_{00}, a_{10}, a_{01}, a_{11}, a_{20}, \cdots$ may be written

$$\begin{aligned} G(t, u) &= a_{00} f_0(t) g_0(u) + a_{10} f_1(t) g_0(u) + a_{01} f_0(t) g_1(u) + \cdots \\ &= \sum_{n=0} \sum_{m=0} a_{nm} f_n(t) g_m(u) \end{aligned} \tag{4}$$

Note that with

$$F_m(t) = \Sigma\, a_{nm} f_n(t)$$

$$G_n(u) = \Sigma\, a_{nm} g_m(u)$$

this may also be written

$$G(t, u) = \sum_{n=0} G_n(u)f_n(t) = \sum_{m=0} F_m(t)g_m(u) \tag{4a}$$

which has the form of (3); treatment of multivariable generating functions in this stepwise fashion is sometimes helpful in finding sets of functions like $f_0(t), f_1(t), \cdots$ which simplify operations and results. The general form of a generating function for any number of variables is sufficiently evident from (4).

A different kind of multivariable generating function has been made the basis of P. A. MacMahon's extensive treatise on combinatorial analysis, already mentioned in the preface. Although little use of it is made in the present work, the idea deserves notice. Take the finite sequence $a_{00}, a_{10}, a_{01}, a_{20}, a_{11}, a_{02}$ and the double power series (two-variable polynomial)

$$A(t, u) = a_{00} + a_{10}t + a_{01}u + a_{20}t^2 + a_{11}tu + a_{02}u^2$$

This is not symmetrical in $t$ and $u$ (for arbitrary coefficients) but the sum $A(t, u) + A(u, t)$ is; in fact

$$A(t, u) + A(u, t) = 2a_{00} + (a_{10} + a_{01})(t + u) + (a_{20} + a_{02})(t^2 + u^2) + 2a_{11}tu \tag{5}$$

and it will be noticed that, on the right symmetric functions, both of the coefficient indices and of the variables $t$ and $u$ appear. Thus (5) is an expansion of a symmetric function in terms of symmetric functions, and is easily extended to any finite or infinite multi-index sequence with, of course, the corresponding number of variables.

With these symmetric generating functions, one function may serve to represent all possible results of a class of combinatorial problems; for example, all numbers of combinations of objects of all possible specifications, rather than those of a single specification as in Chapter 1. However, perhaps because of this very great generality, such functions are not often used. One reason for this is that their algebra is necessarily rather extensive; large numbers of terms must be carried along despite the simplifications achieved by MacMahon's development of the theory of Hammond operators, the natural tool. Another, perhaps more important, reason is that only rarely is such a complete examination of a combinatorial question asked for; the possibilities are so extensive that the mind boggles and is satisfied with a limited, even tiny, selection of these possibilities, and the theory which has all possibilities nicely packaged does not yield these limited answers easily. This must be the justification for not considering this beautiful theory here, but it will be apparent that it has influenced the development in several places (notably the cycles of permutations and the theory of distributions).

Finally, for those readers who are more accustomed to continuous variables, it may be noticed that the analogue of the ordinary generating function of one variable is the infinite integral

$$A(t) = \int_0^\infty a(k)t^k \, dk$$

This is probably more familiar with an exponential kernel $e^{-tk}$ in place of $t^k$ when it becomes the Laplace transform

$$A(t) = \int_0^\infty a(k)e^{-tk} \, dk$$

An expression which contains both the integrals above and the power series of (1) is the Stieltjes integral

$$A(t) = \int_0^\infty a(k) \, dF(t, k)$$

with $t$ a parameter. To obtain (1), $F(t, k)$ is taken as a step function with jumps at the values $k = 0, 1, 2, \cdots$, the jump at $k$ being $t^k$.

*Example 1.* For the sequence defined by

$$a_0 = a_1 = \cdots = a_N = 1$$
$$a_{N+n} = 0, \qquad n > 0$$

the generating function $A(t)$ becomes the polynomial

$$\begin{aligned} A(t) &= 1 + t + t^2 + \cdots + t^N \\ &= (1 - t^{N+1})/(1 - t) \end{aligned}$$

The first two derivatives of $A(t)$ are

$$\begin{aligned} A'(t) &= 1 + 2t + 3t^2 + \cdots + Nt^{N-1} \\ &= (1 - (N+1)t^N + Nt^{N+1})/(1-t)^2 \\ A''(t) &= 2 + 6t + 12t^2 + \cdots + N(N-1)t^{N-2} \\ &= \frac{2 - N(N+1)t^{N-1} + 2(N^2-1)t^N - N(N-1)t^{N+1}}{(1-t)^3} \end{aligned}$$

These results indicate that, for finite sequences, natural expectations are fulfilled: the formal operations on the generating functions produce the generating functions of the allied sequences, namely, the sequence $(n + 1)a_{n+1}$, $n = 0, 1, \cdots, N - 1$ for the first derivative, the sequence $(n + 2)(n + 1)a_n$, $n = 0, 1, \cdots, N - 2$ for the second. It will be noticed also that the use of finite sequences entails a considerable awkwardness in the appearance of terms necessary to truncate the naturally infinite generating functions $(1 - t)^{-1}$, $(1 - t)^{-2}$, $(1 - t)^{-3}$.

The following short table gives generating functions, both ordinary and exponential, for a few simple infinite sequences. The results are all simple and well known, and the table is merely for concreteness.

SOME SIMPLE GENERATING FUNCTIONS

| $a_k$ | $A(t)$ | $E(t)$ |
|---|---|---|
| $a^k$ | $(1 - at)^{-1}$ | $\exp at$ |
| $k$ | $t(1 - t)^{-2}$ | $t \exp t$ |
| $k(k - 1)$ | $2t^2(1 - t)^{-3}$ | $t^2 \exp t$ |
| $k^2$ | $t(t + 1)(1 - t)^{-3}$ | $t(t + 1) \exp t$ |
| $\binom{n}{k}$ | $(1 + t)^n$ | — |
| $n(n - 1) \cdots (n - k + 1)$ | — | $(1 + t)^n$ |

## 2. ELEMENTARY RELATIONS OF ORDINARY GENERATING FUNCTIONS

Write $A(t)$, $B(t)$, $C(t)$ for the generating functions corresponding to the sequences $(a_k)$, $(b_k)$, $(c_k)$. Then, as an immediate consequence of definition $(a)$, in the following pairs of relations, each implies the other.

Sum:

$$a_k = b_k + c_k$$

$$A(t) = B(t) + C(t) \tag{6}$$

Product:

$$a_k = b_k c_0 + b_{k-1} c_1 + \cdots + b_1 c_{k-1} + b_0 c_k \tag{7}$$

$$A(t) = B(t)\, C(t)$$

In (7) the first result is obtained by equating coefficients of like powers of $t$ in the relation $A(t) = B(t)C(t)$; it is clear that a coefficient of $t^k$ can be obtained by matching a term $b_j t^j$ from $B(t)$ with $c_{k-j} t^{k-j}$ from $C(t)$ for $j = 0, 1, \cdots, k$. The sum $b_k c_0 + b_{k-1} c_1 + \cdots + b_0 c_k$ is sometimes called the convolution; more accurately the sequence $(a_k)$ is the convolution of $(b_k)$ and $(c_k)$ when $A(t) = B(t)C(t)$. Equation (7) may also be regarded as a *definition* of the product of the sequences $(a_k)$, $(b_k)$ in the Cauchy algebra.

Note that both sum and product are commutative and associative.

The product (7) has many uses obtained by specialization of one of the functions $B(t)$, $C(t)$. Thus, for example

$$A(t) = B(t)(1 - t) \tag{8}$$

implies

$$a_k = b_k - b_{k-1}$$

or

$$a_k = \Delta b_{k-1} \tag{9}$$

$$b_k = a_k + b_{k-1} = a_k + a_{k-1} + \cdots + a_0 \tag{10}$$

Equation (9) is simply a rewriting of the second equation of (8), which defines $\Delta$, and (10) shows that the generating function inverse of (8), namely $B(t) = A(t)(1 - t)^{-1}$ leads to correct results.

The difference operator $\Delta$ is customarily used in the form $\Delta b_k = b_{k+1} - b_k$; the generating function relation corresponding to $a_k = \Delta b_k$ is

$$A(t) = B(t)(t^{-1} - 1) - b_0 t^{-1}$$

The term $-b_0 t^{-1}$ is added to account for the initial condition.

In probability work, the complementary form of relation (10) is preferred; changing notation, this may be stated as:

$$A(t) = [B(1) - B(t)](1 - t)^{-1} \tag{11}$$

implies

$$a_k = b_{k+1} + b_{k+2} + \cdots$$

and vice versa.

Thus, if $B(t)$ is a polynomial whose coefficients are probabilities, $A(t)$ is the generating function of their cumulations; the usual case here is that $B(1) = 1$.

Iterations of (10) lead to sums of sums; thus, if

$$\begin{aligned} S^0 a_k &= a_k \\ S a_k &= a_k + a_{k-1} + \cdots + a_0 \\ S^2 a_k &= S(a_k + a_{k-1} + \cdots + a_0) \\ &= a_k + 2a_{k-1} + 3a_{k-2} + \cdots + (k + 1)a_0 \end{aligned}$$

then, by induction,

$$\begin{aligned} S^n(a_k) &= S[S^{n-1}(a_k)] \\ &= a_k + na_{k-1} + \cdots + \binom{n+j-1}{j} a_{k-j} + \cdots \\ &\quad + \binom{n+k-1}{k} a_0 \end{aligned}$$

This is equivalent to the generating function relation

$$B(t) = A(t)(1 - t)^{-n}, \qquad b_k = S^n(a_k)$$

The numbers $\binom{n+j-1}{j}$ appearing in $S^n$ have been called figurate numbers (because they enumerate the number of points in certain figures; e.g., for $j = 1$, the figures are triangles); they are also the numbers of $j$-combinations with repetition of $n$ distinct things (see Section 1.3.2).

As a final use of (7), it may be noted that the generating function of the sequence $a'_0, a'_1, \cdots$, inverse to the sequence $a_0, a_1, \cdots$ is defined by

$$A(t)A'(t) = 1$$

so that

$$a_0 a'_0 = 1 \tag{12}$$

$$a_1 a'_0 + a_0 a'_1 = 0$$

$$a_2 a'_0 + a_1 a'_1 + a_0 a'_2 = 0$$

and so on.

*Example 2.* The recurrence relation for combinations with repetition is (1.9), namely,

$$f(n, r) = f(n, r - 1) + f(n - 1, r)$$

Writing

$$f_n(t) = \sum_{r=0}^{\infty} f(n, r)t^r$$

this becomes

$$(1 - t)f_n(t) = f_{n-1}(t)$$

or

$$f_n(t) = (1 - t)^{-1}f_{n-1}(t) = (1 - t)^{-2}f_{n-2}(t) = (1 - t)^{-n}$$

since $f_0(t) = 1$, by convention; or, if we like, $f_1(t) = (1 - t)^{-1}$. Hence, by the binomial expansion,

$$f(n, r) = \binom{n + r - 1}{r}$$

as in (1.10). Many uses of the generating function are equally simple.

The simple results of differentiation noticed below are often useful. Writing $D \equiv d/dt$,

$$DA(t) = \Sigma k a_k t^{k-1} \tag{13}$$

$$D^2A(t) = \Sigma k(k - 1)a_k t^{k-2}$$

$$D^jA(t) = \Sigma (k)_j a_k t^{k-j}, \qquad (k)_j = k(k - 1) \cdots (k - j + 1)$$

To obtain power multipliers of $a_k$, the operator $\theta = tD$ is more convenient; thus,

$$\theta A(t) = \Sigma k a_k t^k \tag{14}$$
$$\theta^2 A(t) = \Sigma k^2 a_k t^k$$
$$\theta^j A(t) = \Sigma k^j a_k t^k$$

Hence, for $P(\theta)$ a polynomial in $\theta$ with constant coefficients

$$P(\theta)A(t) = \Sigma P(k) a_k t^k \tag{15}$$

## 3. SOLUTION OF LINEAR RECURRENCES

The implication of the product of two ordinary generating functions, equations (7), also has a direct use in the solution of linear recurrences. To illustrate, consider a recurrence of the second order.

$$a_{n+2} - \alpha a_{n+1} - \beta a_n = 0, \qquad n = 0, 1, 2, \cdots \tag{16}$$

with $\alpha$, $\beta$ constants independent of $n$. This leaves $a_0$ and $a_1$ undefined, but, if these are regarded as given initial boundary conditions, the system is completed by adding the identities

$$a_0 = a_0$$
$$a_1 - \alpha a_0 = a_1 - \alpha a_0$$

The system of equations is now in the form required by (7), namely

$$a_n b_0 + a_{n-1} b_1 + \cdots + a_0 b_n = c_n$$

with

$$b_0 = 1, \qquad b_1 = -\alpha, \qquad b_2 = -\beta, \qquad b_{n+2} = 0, \qquad n > 0$$
$$c_0 = a_0, \qquad c_1 = a_1 - \alpha a_0, \qquad c_{n+1} = 0, \qquad n > 0$$

Hence,

$$A(t)[1 - \alpha t - \beta t^2] = a_0 + (a_1 - \alpha a_0)t \tag{17}$$

which, of course, could also have been obtained by multiplying the $n$th equation of the complete system through by $t^n$ and summing. An explicit expression follows by expansion in partial fractions; thus, if

$$1 - \alpha t - \beta t^2 = (1 - s_1 t)(1 - s_2 t)$$

then

$$A(t) = \frac{1}{s_1 - s_2}\left(\frac{a_1 - \alpha a_0 + s_1 a_0}{1 - s_1 t} - \frac{a_1 - \alpha a_0 + s_2 a_0}{1 - s_2 t}\right)$$

and, by equating coefficients of $t^n$,

$$a_n = (a_1 - \alpha a_0)\frac{s_1^n - s_2^n}{s_1 - s_2} + a_0\frac{s_1^{n+1} - s_2^{n+1}}{s_1 - s_2} \tag{18}$$

It may be verified that (18) reduces to $a_0$ and $a_1$ for $n = 0$ and 1 respectively, and also satisfies recurrence (16).

## 4. EXPONENTIAL GENERATING FUNCTIONS

Writing $E(t)$, $F(t)$, and $G(t)$ for the exponential generating functions of the sequences $(a_k)$, $(b_k)$, and $(c_k)$, the basic relations for sum and product of sequences are given by

Sum:
$$a_k = b_k + c_k \tag{19}$$
$$E(t) = F(t) + G(t)$$

Product:
$$a_k = b_kc_0 + kb_{k-1}c_1 + \cdots + \binom{k}{j} b_{k-j}c_j + \cdots + b_0c_k \tag{20}$$
$$E(t) = F(t)G(t)$$

Equations (19) are formally the same as the similar equations for the ordinary generating functions. Equations (20) differ from their correspondents, equations (7), through the presence of binomial coefficients, which suggests a basic device of the Blissard calculus.

This is as follows. A sequence $a_0, a_1, \cdots$ may be replaced by $a^0$, $a^1, \cdots$, with the exponents treated as powers during all formal operations, and only restored as indexes when operations are completed; note that $a^0$ is not necessarily equal to 1. Thus, with this device, the first equation of (20) is written as

$$a_k = (b + c)^k, \qquad b^n \equiv b_n, \qquad c^n \equiv c_n \tag{20a}$$

the additional qualifiers, $b^n \equiv b_n$, $c^n \equiv c_n$, serving as reminders. Note that

$$a_0 = b_0c_0$$

which is equal to unity only when both $b_0$ and $c_0$ are unity.

Again, as with equation (7), either (20) or (20*a*) is a *definition* of the product sequence in this algebra.

The Blissard device mentioned above is, of course, suggested by the fact that, with it, exponential generating functions look like exponential functions; thus

$$E(t) = \exp at, \qquad a^n \equiv a_n$$

and with similar expressions for $F(t)$ and $G(t)$, the second equation of (20) becomes

$$\exp at = (\exp bt)(\exp ct) = \exp (b + c)t$$

Equating coefficients of $t^n$ gives (20a).

It may also be noticed that, with a prime denoting a derivative,

$$\begin{aligned} E'(t) &= a_1 + a_2 t + a_3 t^2/2! + \cdots \\ &= a \exp at, \qquad a^n \equiv a_n \end{aligned}$$

The extension of (20) or (20a) to many variables results in

$$\begin{aligned} a_0 &= b_0 c_0 \cdots \\ a_n &= (b + c + \cdots)^n, \qquad b^n \equiv b_n, \qquad c^n \equiv c_n, \cdots \\ &= \Sigma \frac{n!}{p!\,q!\cdots} b_p c_q \cdots, \qquad p + q + \cdots = n \end{aligned} \tag{21}$$

the last form, of course, being an expression of the multinomial theorem.

Like sequences are treated exactly as unlike sequences; thus,

$$(a + a)^n = \sum_{k=0}^{n} \binom{n}{k} a_{n-k} a_k \tag{22}$$

and not

$$(a + a)^n = \Sigma \binom{n}{k} a^{n-k} a^k = \Sigma \binom{n}{k} a_n = 2^n a_n$$

To complete the algebra, note that the sequence $(a'_n)$ inverse to a given sequence $(a_n)$ is defined by

$$a_0 a'_0 = 1 \tag{23}$$

$$(a + a')^n = 0, \qquad n > 0, \qquad a^n \equiv a_n, \qquad (a')^n \equiv a'_n$$

which is equivalent to the generating function relation:

$$\exp (a + a')t = 1, \qquad a^n \equiv a_n, \qquad (a')^n \equiv a'_n$$

Equations (23) may be solved for variables of either of the sequences in terms of those of the other; a concise expression of this solution appears later in the chapter, but it may be noted now that

$$\begin{aligned} a'_0 &= 1/a_0 & a'_2 &= -a_2/a_0^2 + 2a_1^2/a_0^3 \\ a'_1 &= -a_1/a_0^2 & a'_3 &= -a_3/a_0^2 + 6a_2 a_1/a_0^3 - 6a_1^3/a_0^4 \end{aligned}$$

*Example 3.* (a) If $\alpha$ is an ordinary (non-symbolic) variable, then each of the equations

$$\exp at = (\exp bt)(\exp \alpha t), \qquad a^n \equiv a_n, \qquad b^n \equiv b_n$$

$$(\exp at)(\exp - \alpha t) = \exp bt, \qquad a^n \equiv a_n, \qquad b^n \equiv b_n$$

implies the other. Hence, each of the equations

$$a_n = (b + \alpha)^n, \qquad b^n \equiv b_n$$
$$b_n = (a - \alpha)^n, \qquad a^n \equiv a_n$$

implies the other.

(*b*) If $\alpha$ is replaced by $c$, the symbolic variable for the sequence $c_0, c_1, \cdots$, and if $c'$ is the corresponding variable for its inverse sequence, the equations above become

$$\exp at = (\exp bt)(\exp ct), \qquad a^n \equiv a_n, \qquad b^n \equiv b_n, \qquad c^n \equiv c_n$$
$$(\exp at)(\exp c't) = \exp bt, \qquad a^n \equiv a_n, \qquad b^n \equiv b_n, \qquad (c')^n \equiv c'_n$$
$$a_n = (b + c)^n, \qquad b^n \equiv b_n, \qquad c^n \equiv c_n$$
$$b_n = (a + c')^n, \qquad a^n \equiv a_n, \qquad (c')^n \equiv c'_n$$

it should be noticed that the identifications $c_0 = 1$, $c_n = \alpha^n$, $n = 1, 2, \cdots$ turn $c$ into the ordinary variable $\alpha$; hence it follows from part (*a*) of this example that $c'_0 = 1$, $c'_n = (-\alpha)^n$, $n = 1, 2, \cdots$, which agrees with (23).

Finally, it may be useful to notice that with $D = d/dt$,

$$D^n E(t) = \sum_{k=0} a_{k+n} t^k/k!$$
$$= a^n \exp at, \qquad a^k \equiv a_k$$

## 5. RELATION OF ORDINARY AND EXPONENTIAL GENERATING FUNCTIONS

If $A(t)$ and $E(t)$ are ordinary and exponential generating functions of the sequence $a_0, a_1, \cdots$, then formally

$$A(t) = \int_0^\infty e^{-s} E(ts)\, ds \tag{24}$$

This is just the relation for Borel summation of divergent series. It is derived by noticing that

$$k! = \int_0^\infty e^{-s} s^k\, ds$$

(the Euler integral for the Gamma function). Hence

$$A(t) = \sum_{k=0} a_k t^k$$
$$= \Sigma\, a_k (t^k/k!) \int_0^\infty e^{-s} s^k\, ds$$
$$= \int_0^\infty e^{-s}\, ds\, \Sigma\, a_k (st)^k/k!$$

*Example 4.* (*a*) For the sequence $1 = a_0 = a_1 = \cdots$,

$$A(t) = 1 + t + t^2 + \cdots \quad = (1 - t)^{-1}$$

$$E(t) = 1 + t + t^2/2! + \cdots = e^t$$

and

$$(1 - t)^{-1} = \int_0^\infty e^{-s} e^{ts}\, ds$$

(*b*) For the sequence $a_0 = a_1 = \cdots = a_{j-1} = 0$, $a_k = (k)_j$, $k \geq j$,

$$A(t) = j!\, t^j + \frac{(j+1)!\, t^{j+1}}{1!} + \cdots \frac{k!\, t^k}{(k-j)!} + \cdots = \frac{t^j j!}{(1-t)^{j+1}}$$

$$E(t) = t^j + \frac{t^{j+1}}{1!} + \cdots \frac{t^{j+k}}{k!} + \cdots = t^j e^t$$

and

$$t^j j! (1 - t)^{-j-1} = \int_0^\infty e^{-s} (st)^j e^{st}\, ds$$

## 6. MOMENT GENERATING FUNCTIONS

If $p_0, p_1, p_2, \cdots$ is a sequence of probabilities (probability distribution) so that $0 \leq p_j \leq 1, j = 0, 1, \cdots$, its *ordinary* moments are defined by

$$m_0 = p_0 + p_1 + p_2 + \cdots$$

$$m_1 = p_1 + 2p_2 + \cdots$$

and

$$\begin{aligned} m_k &= p_1 + 2^k p_2 + 3^k p_3 + \cdots \\ &= \sum_{j=0} j^k p_j \end{aligned} \tag{25}$$

The *factorial* moments are defined by

$$(m)_k = \sum_{j=0} (j)_k p_j, \qquad (j)_k = j(j-1) \cdots (j-k+1) \tag{26}$$

and the closely related *binomial* moments by

$$B_k = \Sigma \binom{j}{k} p_j = (m)_k / k! \tag{27}$$

The notation $(m)_k$ is adopted for factorial moments because

$$(m)_k = m(m-1) \cdots (m-k+1), \qquad m^j \equiv m_j \tag{28}$$

where $m_j$ is an ordinary moment. Anticipating the next section, it may be noted that the numbers appearing when (28) is fully developed are

Stirling numbers of the first kind, whereas those in the inverse expression of ordinary moments by factorial moments are Stirling numbers of the second kind.

The *central* moments are defined by

$$M_k = (m - m_1)^k, \qquad m^k \equiv m_k \tag{29}$$

$$= \Sigma \binom{k}{j} m_{k-j}(-m_1)^j$$

In particular the second central moment $M_2 = m_2 - m_1^2$ is known as the variance.

The generating functions for these moments are all interrelated. If $P(t)$ is the ordinary generating function for the probabilities, that is,

$$P(t) = p_0 + p_1 t + p_2 t^2 + \cdots$$

then

$$m(t) = \exp mt = \sum_{k=0}^{\infty} m_k t^k/k! = P(e^t) \tag{30}$$

$$\exp (m)t = \sum_{k=0}^{\infty} (m)_k t^k/k! = P(1 + t)$$

$$B(t) = B_0 + B_1 t + \cdots = \exp (m)t = P(1 + t)$$

and

$$m(t) = B(e^t - 1) \tag{31}$$

These relations are all obtained by formal operations; thus, for the first of (30)

$$P(e^t) = \sum_{j=0} p_j e^{tj}$$

$$= \sum_{j=0} p_j \sum_{k=0} (tj)^k/k!$$

$$= \sum_{k=0} t^k/k! \sum_{j=0} p_j j^k$$

It may be noticed that the last of equations (30) may be written

$$B(t) = (1 + t)^m, \qquad m^k \equiv m_k \tag{30a}$$

whence

$$B(e^t - 1) = (e^t)^m = \exp mt, \qquad m^k \equiv m_k$$

which is equation (31).

The second of equations (30) is equivalent to

$$P(t) = \exp (m)(t - 1)$$

$$= (\exp - (m))(\exp (m)t), \qquad (m)^k \equiv (m)_k$$

and hence to

$$p_j = \sum_{k=0} (-1)^k \frac{(m)_{k+j}}{j!\,k!} \tag{32}$$

$$= \sum_{k=0} (-1)^k \binom{k+j}{k} B_{j+k}$$

an important result closely related to the principle of inclusion and exclusion, which will be examined in the next chapter; this is because in many problems the factorial moments are easier to determine than the probabilities which are then given by (32). It should be observed that the second form of (32) is a reciprocal relation to

$$B_k = \Sigma \binom{j}{k} p_j$$

the definition equation for binomial moments.

The generating function for central moments is

$$\begin{aligned} M(t) = \exp Mt &= \Sigma M_k t^k/k! \\ &= \exp (m - m_1)t \\ &= (\exp -m_1 t)m(t) \end{aligned} \tag{33}$$

There are, of course, corresponding multivariable probability and moment generating functions for multi-index probability sequences, which are interelated in a similar manner. It may be noted that the *covariance* of a two-index distribution is given by

$$m_{11} - m_{10}m_{01}$$

with

$$m_{rs} = \Sigma j^r k^s p_{jk}$$

## 7. STIRLING NUMBERS

As noted above, Stirling numbers (named for their discoverer) are those numbers appearing in the sums relating powers of a variable to its factorials, and vice versa. Because in the finite difference calculus factorials have the same pre-eminence that powers have in the differential calculus, the numbers form a part of the natural bridge between these two calculi and are continually being rediscovered (almost as often as their next of kin, the Bernoulli numbers).

They are defined as follows. If

$$(t)_0 = t^0 = s(0, 0) = S(0, 0) = 1$$

$$(t)_n = t(t-1)\cdots(t-n+1)$$

$$= \sum_0^n s(n, k)t^k, \qquad n > 0 \tag{34}$$

$$t^n = \sum_0^n S(n, k)(t)_k, \qquad n > 0 \tag{35}$$

then $s(n, k)$, $S(n, k)$ are Stirling numbers of the first and second kinds respectively. Note that both kinds of numbers are non-zero only for $k = 1, 2, \cdots n$, $n > 0$, and that $(t)_n$ is an ordinary generating function for $s(n, k)$ whereas $t^n$ is a new kind of generating function with $f_k(t)$ of equation (3) equal to $(t)_k$ (it will be seen that this set of functions is linearly independent). For given $n$, or given $k$, the numbers of the first kind $s(n, k)$ alternate in sign; indeed, since

$$(-t)_n = (-1)^n t(t+1)\cdots(t+n-1)$$

it follows at once from (34) that $(-1)^{n+k}s(n, k)$ is always positive. Sometimes, as in Chapter 4, these positive numbers, which are generated by rising factorials, are more convenient.

Recurrence relations for the numbers arise from the following simple recurrence for factorials:

$$(t)_{n+1} = (t-n)(t)_n$$

Used with (34), this implies

$$s(n+1, k) = s(n, k-1) - ns(n, k) \tag{36}$$

whereas for (35)

$$t^{n+1} = \Sigma S(n+1, k)(t)_k$$

$$= t\Sigma S(n, k)(t)_k$$

$$= \Sigma S(n, k)[(t)_{k+1} + k(t)_k]$$

and

$$S(n+1, k) = S(n, k-1) + kS(n, k) \tag{37}$$

These may be used to determine the numbers shown in Tables 1 and 2 at the end of this chapter, in which $k = 1(1)n$, $n = 1(1)10$.

The numbers $S(n, k)$ are related to the numbers $\Delta^k 0^n$ of Example 1.12 by means of

$$k!\, S(n, k) = \Delta^k 0^n \tag{38}$$

This may be found with great brevity by the symbolic operations

$$t^n = E^t 0^n = (1 + \Delta)^t 0^n$$
$$= \Sigma \binom{t}{k} \Delta^k 0^n$$

and, of course, may be verified otherwise.

Finally, inserting (34) into (35), or vice versa, shows that

$$\Sigma S(n, k)s(k, m) = \delta_{nm} \tag{39}$$

with $\delta_{nm}$ a Kronecker delta: $\delta_{nn} = 1$, $\delta_{nm} = 0$, $n \neq m$, and the sum over all values of $k$ for which $S(n, k)$ and $s(k, m)$ are non-zero. This shows that each of the relations

$$a_n = \Sigma s(n, k)b_k \tag{40}$$
$$b_n = \Sigma S(n, k)a_k$$

implies the other. An example of equations (40) already mentioned is that in which $a_n = (m)_n$, the $n$th factorial moment of a probability distribution, and $b_n = m_n$, the $n$th ordinary moment; the two equations of (40) then are

$$(m)_n = \Sigma s(n, k)m_k, \qquad m_n = \Sigma S(n, k)(m)_k$$

If $S$ is the infinite matrix $(S(n, k))$, that is, the matrix with $S(n, k)$ the entry in row $n$ and column $k$, $n = 0, 1, 2, \cdots, k = 0, 1, 2, \cdots$, and if $s$ is the similar matrix for $s(n, k)$, then equation (39) is equivalent to the matrix equation

$$Ss = I$$

with $I$ the unit infinite matrix, $(\delta_{nk})$. Hence $s = S^{-1}$ is the inverse of $S$, and vice versa. This suggests a doubly infinite family of Stirling numbers (introduced by E. T. Bell, **5**) defined by the matrix relations

$$S^1 = S \cdot I = S$$
$$S^r = S \cdot S^{r-1}$$

the typical element of $S^r$ being $S(n, k, r)$. The general matrix equation is then the same as

$$S(n, k, r) = \Sigma S(n, j, r - 1)S(j, k, 1)$$

## 8. DERIVATIVES OF COMPOSITE FUNCTIONS

A composite function is a function of a function. It may be put in the standard form of an exponential (or ordinary) generating function through its derivatives. Expressed in terms of the derivatives of the component functions, the latter form a set of polynomials, the Bell polynomials, **4**, which are an important feature of many combinatorial and statistical problems.

Write

$$A(t) = f[g(t)] \tag{41}$$

and, with $D_t = d/dt$, $D_u = d/du$,

$$D_t^n A(t) = A_n, \qquad [D_u^n f(u)]_{u=g(t)} = f_n, \qquad D_t^n g(t) = g_n$$

Then, by successive differentiation of (41)

$$\begin{aligned} A_1 &= f_1 g_1 \\ A_2 &= f_1 g_2 + f_2 g_1^2 \\ A_3 &= f_1 g_3 + 3 f_2 g_2 g_1 + f_3 g_1^3 \end{aligned}$$

The general form may be written

$$\begin{aligned} A_n &= f_1 A_{n1} + f_2 A_{n2} + \cdots + f_n A_{nn} \\ &= \sum_{k=1}^{n} f_k A_{n,k}(g_1, g_2, \cdots, g_n) \end{aligned} \tag{42}$$

Note that the coefficients $A_{n,k}$ depend only on the derivatives $g_1, g_2, \cdots$, as indicated by the expanded notation of the second line, and not on the $f_k$. Hence they may be determined by a special choice of $f$; a convenient one is $f(u) = \exp au$, $a$ being a constant, which entails

$$f_k = a^k \exp ag, \qquad g = g(t)$$

and

$$\begin{aligned} e^{-ag} D_t^n e^{ag} &= \Sigma A_{n,k}(g_1, g_2, \cdots, g_n) a^k \\ &= A_n(a; g_1, g_2, \cdots, g_n) \end{aligned} \tag{43}$$

with the second line a definition to abbreviate the first. In this notation equation (42) becomes

$$A_n = A_n(f; g_1, g_2, \cdots, g_n), \qquad f^k \equiv f_k \tag{42a}$$

with

$$A_0 = f_0 = A(t)$$

Equation (43) completely determines the polynomials $A_n(a; g_1, \cdots, g_n)$ and through (42*a*) the derivatives $A_n$. It may be noted here that in Bell's notation

$$\begin{aligned} A_n(1; y_1, y_2, \cdots, y_n) &= Y_n(y_1, y_2, \cdots, y_n) \\ &= e^{-y} D_x^n e^y, \qquad y \equiv y(x) \end{aligned}$$

To obtain an explicit formula, note first that, abbreviating

$$A_n(a; g_1, \cdots, g_n)$$

to $A_n(a)$, and using the Leibniz formula for differentiation of a product

$$\begin{aligned}
A_{n+1}(a) &= e^{-ag} D^n(De^{ag}) \\
&= e^{-ag}a\, D^n(g_1 e^{ag}) \\
&= a \sum_{k=0}^{n} \binom{n}{k} (e^{-ag} D^{n-k} e^{ag})\, D^k g_1 \\
&= a \sum_{k=0}^{n} \binom{n}{k} A_{n-k}(a) g_{k+1} \\
&= ag(A(a) + g)^n, \qquad (A(a))^k \equiv A_k(a), \qquad g^k \equiv g_k
\end{aligned} \tag{44}$$

Instances of (44) with $A_0(a) = 1$ are

$$\begin{aligned}
A_1(a) &= ag_1 \\
A_2(a) &= ag_2 + ag_1 A_1(a) = ag_2 + a^2 g_1^2 \\
A_3(a) &= ag_3 + 2ag_2 A_1(a) + ag_1 A_2(a) \\
&= ag_3 + 3a^2 g_2 g_1 + a^3 g_1^3
\end{aligned}$$

which agree with the results preceding (42).

Next, (44) implies the exponential generating function relation

$$\begin{aligned}
\exp uA(a) &= \sum_{n=0}^{\infty} A_n(a) u^n/n! \\
&= \exp a[ug_1 + u^2 g_2/2! + \cdots] \\
&= \exp aG(u)
\end{aligned} \tag{45}$$

with

$$G(u) = \exp(ug) - g_0, \qquad g^n \equiv g_n$$

Differentiation of (45) and equating coefficients of $u^n$ gives (44).

Finally, expanding (44), using the multinomial theorem, and equating coefficients of $u^n$ gives the explicit formula

$$A_n(a) = \sum \frac{a^k n!}{k_1! \cdots k_n!} \left(\frac{g_1}{1!}\right)^{k_1} \cdots \left(\frac{g_n}{n!}\right)^{k_n} \tag{46}$$

with $k = k_1 + k_2 + \cdots + k_n$ and the sum over all solutions in non-negative integers of $k_1 + 2k_2 + \cdots + nk_n = n$, or over all partitions of $n$ (anticipating Chapter 6). Then, by (42$a$),

$$A_n = A_n(f) = \sum \frac{n!\, f_k}{k_1! \cdots k_n!} \left(\frac{g_1}{1!}\right)^{k_1} \cdots \left(\frac{g_n}{n!}\right)^{k_n} \tag{45a}$$

which is known as di Bruno's formula. Table 3 at the end of this chapter shows these polynomials, in Bell's notation (slightly modified), for $n = 1(1)8$.

It is worth noting that, if $A(t)$ has a Taylor's series expansion, then

$$A(t + u) = \exp uA(f), \qquad A^n(f) \equiv A_n(f)$$

so that, if

$$A_n^0 = A_n(f) \qquad \text{for} \qquad t = 0$$
$$A(t) = \exp uA^0, \qquad (A^0)^n \equiv A_n^0$$

The polynomials have a direct use in statistics in relating the cumulants (semi-invariants) to ordinary moments; this is of general interest because it raises the problem of inverting the polynomials.

The cumulant (exponential) generating function

$$L(t) = \lambda_1 t + \lambda_2 t^2 + \cdots + \lambda_n t^n/n! \cdots$$

is usually defined by the relation

$$\exp mt = \exp L(t), \qquad m^n \equiv m_n$$

where $m_n$ is the $n$th ordinary moment. It follows at once from (45) with $a = 1$ that

$$m_n = A_n(1; \lambda_1, \cdots, \lambda_n) = Y_n(\lambda_1, \lambda_2, \cdots, \lambda_n) \tag{47}$$

and from (44) that

$$m_{n+1} = \lambda(m + \lambda)^n, \qquad m^n \equiv m_n, \qquad \lambda^n \equiv \lambda_n \tag{48}$$

From either of these the following inverse relations are found readily

$$\lambda_1 = m_1 \qquad \lambda_3 = m_3 - 3m_2m_1 + 2m_1^3$$
$$\lambda_2 = m_2 - m_1^2 \quad \lambda_4 = m_4 - 4m_3m_1 - 3m_2^2 + 12m_2m_1^2 - 6m_1^4$$

The inverse relation to (45) is

$$aG(u) = \log(\exp uA(a)) \tag{49}$$

which is in the form of (41) with $f(u) = \log u$; hence

$$ag_n = A_n(f; A_1(a), \cdots, A_n(a)), \qquad f^k \equiv f_k \equiv (-1)^{k-1}(k-1)! \tag{50}$$

is the required set of inverse relations. Thus

$$\lambda_n = A_n(f; m_1, \cdots, m_n), \qquad f^k \equiv f_k \equiv (-1)^{k-1}(k-1)! \tag{51}$$
$$= Y_n(fm_1, \cdots, fm_n), \qquad f^k \equiv f_k \equiv (-1)^{k-1}(k-1)!$$

is the inverse of (47), and will be found to agree with the specific formulas given above.

Other properties of the Bell polynomials and examples of their use appear in the problems.

## REFERENCES

1. E. T. Bell, Euler algebra, *Trans. Amer. Math. Soc.*, vol. 25 (1923), pp. 135–154.
2. ———, *Algebraic Arithmetic*, New York, 1927.
3. ———, Postulational bases for the umbral calculus, *Amer. Journal of Math.*, vol. 62 (1940), pp. 717–724.
4. ———, Exponential polynomials, *Annals of Math.*, vol 35 (1934), pp. 258–277.
5. ———, Generalized Stirling transforms of sequences, *Amer. Journal of Math.*, vol. 61 (1939), pp. 89–101.
6. G. H. Hardy and E. M. Wright, *An Introduction to the Theory of Numbers*, Oxford, 1938.
7. E. Lucas, *Théorie des nombres*, Paris, 1891, chapter 13.
8. P. A. MacMahon, *Combinatory Analysis* (vol. I), London, 1915; (vol. II), London, 1916.

## PROBLEMS

1. If $A(t) = a_0 + a_1 t + \cdots + a_N t^N$, show that

$$a_k = \frac{1}{2\pi}\int_{-\pi}^{\pi} e^{-iku} A(e^{iu})\, du \qquad \text{(Laplace)}$$

2. (*a*) If $a_n(x)$ is defined by

$$a_n(x) = (1 - x)^{n+1}\sum_0^\infty k^n x^k$$

show that $(D = d/dx)$

$$a_n(x) = nxa_{n-1}(x) + x(1 - x)Da_{n-1}(x)$$

and verify the values

$$a_0(x) = 1 \qquad a_2(x) = x + x^2$$
$$a_1(x) = x \qquad a_3(x) = x + 4x^2 + x^3$$

(*b*) If $a_n(x) = a_{n,1}x + a_{n,2}x^2 + \cdots, n > 0$, show that

$$a_{n,k} = ka_{n-1,k} + (n - k + 1)a_{n-1,k-1},\ n > 0$$

and verify the values

$$a_{n,1} = 1,$$
$$a_{n,2} = 2^n - (n + 1),$$
$$a_{n,3} = 3^n - (n + 1)2^n + \binom{n+1}{2}$$

(*c*) Use the definition of $a_n(x)$ to show that its exponential generating function

$$a(x, t) = a_0(x) + a_1(x)t + a_2(x)t^2/2! + \cdots$$

is given by

$$a(x, t) = (1 - x)/(1 - xe^{t(1-x)})$$

Using subscripts to denote partial derivatives, derive

$$(1 - xt)a_t(x, t) = xa(x, t) + x(1 - x)a_x(x, t)$$

from the recurrence relation of part (*a*), and verify that the expression above satisfies this partial-differential equation.

(*d*) Define a new set of functions $A_n(x)$ by the relations:

$$A_0(x) = a_0(x) = 1, \qquad xA_n(x) = a_n(x)$$

and derive the results

$$\begin{aligned} A(x, t) &= A_0(x) + A_1(x)t + A_2(x)t^2/2! + \cdots \\ &= (1 - x)/(e^{t(x-1)} - x) \\ (1 - xt)A_t(x, t) &= A(x, t) + x(1 - x)A_x(x, t) \end{aligned}$$

(*e*) Define $H_n(x)$ by $H_n(x) = A_n(x)/(x - 1)^n$ and derive the results:

$$\begin{aligned} H(x, t) &= H_0(x) + H_1(x)t + H_2(x)t^2/2! + \cdots \\ &= (1 - x)/(e^t - x) \qquad \text{(Euler)} \\ (H + 1)^n &= xH_n + (1 - x)\,\delta_{n0}, \qquad H^n \equiv H_n = H_n(x) \end{aligned}$$

where $\delta_{n0}$ is a Kronecker delta.

Because of these results, the numbers $a_{n,k}$, or, what is the same thing, the numbers $A_{n,k}$ defined by

$$A_n(x) = A_{n,1} + A_{n,2}x + A_{n,3}x^2 + \cdots$$

are called Eulerian numbers. A table appears in Chapter 8.

3. If

$$a_k = \binom{k}{0} + \binom{k+1}{2} + \cdots + \binom{k+j}{2j} + \cdots + \binom{2k}{2k}, \qquad k \geq 0$$

$$b_k = \binom{k}{1} + \binom{k+1}{3} + \cdots + \binom{k+j}{1+2j} + \cdots + \binom{2k-1}{2k-1}, \qquad k > 0$$

show that

$$\begin{aligned} a_{k+1} &= a_k + b_{k+1} \\ b_{k+1} &= a_k + b_k \end{aligned}$$

Hence show that their ordinary generating functions

$$\begin{aligned} A(t) &= a_0 + a_1t + \cdots + a_nt^n + \cdots \\ B(t) &= b_0 + b_1t + \cdots + b_nt^n + \cdots \end{aligned}$$

(note that $a_0 = 1$, $b_0 = 0$) are related by

$$\begin{aligned} A(t) - 1 &= tA(t) + B(t) \\ B(t) &= tA(t) + tB(t) \end{aligned}$$

so that

$$A(t) = (1 - t)[(1 - t)^2 - t]^{-1}$$
$$B(t) = t[(1 - t)^2 - t]^{-1}$$

4. Similarly, if $b_0 = c_0 = c_1 = 0$, and

$$a_k = \binom{k}{0} + \binom{k+1}{3} + \cdots + \binom{k+j}{3j} + \cdots, \qquad k \geq 0$$

$$b_k = \binom{k}{1} + \binom{k+1}{4} + \cdots + \binom{k+j}{1+3j} + \cdots, \qquad k > 0$$

$$c_k = \binom{k}{2} + \binom{k+1}{5} + \cdots + \binom{k+j}{2+3j} + \cdots, \qquad k > 1$$

(all three series are finite, with upper limits provided naturally by the binomial coefficients) show that

$$\begin{aligned} a_{k+1} &= a_k \qquad\quad + c_{k+1} \\ b_{k+1} &= a_k + b_k \\ c_{k+1} &= \qquad\quad b_k + c_k \end{aligned}$$

and that the corresponding relations for generating functions are

$$\begin{aligned} A(t) - 1 &= tA(t) \qquad\qquad + C(t) \\ B(t) \quad &= tA(t) + tB(t) \\ C(t) \quad &= \qquad\qquad tB(t) + tC(t) \end{aligned}$$

and

$$\begin{aligned} A(t) &= (1 - t)^2[(1 - t)^3 - t^2]^{-1} \\ B(t) &= t(1 - t)[(1 - t)^3 - t^2]^{-1} \\ C(t) &= t^2[(1 - t)^3 - t^2]^{-1} \end{aligned}$$

5. Problems 3 and 4 may be generalized to consider finite sums

$$a_k = \binom{k}{b} + \binom{k+a}{b+c} + \binom{k+2a}{b+2c} + \cdots, \qquad a, b < c$$

Show that the generating function of the sum indicated, namely,

$$A(t; a, b, c) = a_0 + a_1 t + a_2 t^2 + \cdots$$

is

$$A(t; a, b, c) = t^b(1 - t)^{c-1-b}[(1 - t)^c - t^{c-a}]^{-1}, \qquad a, b < c$$

6. (*a*) Take $\alpha = e^{i2\pi/3}$, $i = \sqrt{-1}$, ($\alpha^3 = 1$, $1 + \alpha + \alpha^2 = 0$); show that

$$\tfrac{1}{3}[2^k + (1 + \alpha)^k \alpha^{2b} + (1 + \alpha^2)^k \alpha^b]$$
$$= \binom{k}{b} + \binom{k}{b+3} + \binom{k}{b+6} + \cdots, \qquad b = 0, 1, 2$$

(*b*) Obtain the same results by expanding

$$A(t; 0, b, 3) = t^b(1 - t)^{2-b}[(1 - t)^3 - t^3]^{-1}$$

of Problem 5 in partial fractions.

(*c*) Simplify these results to the following

$$\frac{1}{3}\left(2^k + 2\cos\frac{k\pi}{3}\right) = \binom{k}{0} + \binom{k}{3} + \binom{k}{6} + \cdots$$

$$\frac{1}{3}\left(2^k + 2\cos\frac{(k-2)\pi}{3}\right) = \binom{k}{1} + \binom{k}{4} + \binom{k}{7} + \cdots$$

$$\frac{1}{3}\left(2^k + 2\cos\frac{(k-4)\pi}{3}\right) = \binom{k}{2} + \binom{k}{5} + \binom{k}{8} + \cdots$$

7. Similarly with $\alpha = \exp i2\pi/c$, $i = \sqrt{-1}$, and $c$ a positive integer, show that

$$\frac{1}{c}\sum_{j=0}^{c-1}(1+\alpha^j)^k\alpha^{-bj} = \binom{k}{b} + \binom{k}{b+c} + \binom{k}{b+2c} + \cdots, \qquad b < c$$

$$= \frac{1}{c}\sum_{j=0}^{c-1}\left(2\cos\frac{j\pi}{c}\right)^k \cos\left(\frac{k-2b)j\pi}{c}\right)$$

8. If the (binomial probability) generating function $B(t)$ is defined by

$$B(t) = \sum_0^n b_k t^k = (q+pt)^n, \qquad p + q = 1$$

show that

$$b_k = \binom{n}{k} q^{n-k}p^k$$

(Note that $b_k$ is a function of $n$ and $p$ as well as of $k$.)

9. *Continuation.* (*a*) Show that the corresponding factorial moment generating function is

$$\exp t(m) = (1+pt)^n$$

and

$$(m)_k = (n)_k p^k$$

(*b*) Show that, with $S(k, j)$ a Stirling number (second kind)

$$m_k \equiv m_k(n, p) = \Sigma\, S(k, j)(n)_j p^j$$

10. *Continuation.* (*a*) The exponential generating function for the ordinary moments $m_k(n, p) \equiv m_k(n)$ is

$$m(n, t) = \exp tm(n) = B(e^t) = (q + pe^t)^n$$

with $[m(n)]^k \equiv m_k(n)$. With $D = d/dt$, derive

$$\begin{aligned} D\exp tm(n) &= npe^t \exp tm(n-1) \\ &= n\exp tm(n) - qn\exp tm(n-1) \end{aligned}$$

and from these the recurrences

$$\begin{aligned} m_{k+1}(n) &= np[m(n-1)+1]^k, \qquad [m(n-1)]^j \equiv m_j(n-1) \\ &= nm_k(n) - qnm_k(n-1) \end{aligned}$$

Note that $np = m_1(n)$.

(*b*) Verify the following instances of the first of these recurrences by comparison with Problem 9(*b*), or otherwise

$$m_1(n) = np$$
$$m_2(n) = np[(n-1)p+1]$$
$$m_3(n) = np[(n-1)(n-2)p^2 + 3(n-1)p + 1]$$

11. *Continuation.* Show that ($D_p = d/dp$, $D_t = d/dt$)

$$pq\, D_p \exp tm(n) = D_t \exp tm(n) - np \exp tm(n)$$

and hence that

$$m_{k+1}(n) = npm_k(n) + pq\, D_p m_k(n)$$

12. *Continuation.* With $M_k(n, p) \equiv M_k(n)$, the $k$th central moment of the binomial distribution, and $\exp tM(n)$ its generating function, show that, with $D = d/dt$,

$$\begin{aligned} D \exp tM(n) &= -np \exp tM(n) + np \exp t[M(n-1) + q] \\ &= nq \exp tM(n) - nq \exp t[M(n-1) - p] \end{aligned}$$

and hence that

$$\begin{aligned} M_{k+1}(n) &= -npM_k(n) + np[M(n-1) + q]^k \\ &= nqM_k(n) - nq[M(n-1) - p]^k \end{aligned}$$

with, of course, in both $[M(n-1)]^j \equiv M_j(n-1)$. Verify the particular instances

$$M_0(n) = 1 \qquad M_2(n) = npq$$
$$M_1(n) = 0 \qquad M_3(n) = npq(q-p)$$

13. (*a*) From the Stirling number recurrence

$$s(n+1, k) + ns(n, k) = s(n, k-1)$$

show that the exponential generating function

$$y_k(t) = \sum_0^\infty s(n, k)t^n/n!, = \exp ts(, k), \qquad s(, k)^n = s(n, k)$$

satisfies the relation

$$(1+t)\frac{dy_k(t)}{dt} = y_{k-1}(t)$$

and hence verify that

$$y_k(t) = [\log(1+t)]^k/k!$$

(*b*) From the differential equation above, derive

$$s(n+1, k) = \sum_{j=0}^{n} \binom{n}{j} s(n-j, k-1)(-1)^j j!$$

and verify by iteration of equation (36).

14. (*a*) Show similarly that the exponential generating function

$$Y_k(t) = \sum_0^\infty S(n, k)t^n/n!, = \exp tS(, k), \qquad S(, k)^n = S(n, k)$$

has the differential recurrence

$$\left(\frac{d}{dt} - k\right) Y_k(t) = Y_{k-1}(t)$$

and verify that

$$Y_1 = e^t - 1$$
$$Y_k = (e^t - 1)^k/k!$$

(*b*) Show that

$$S(n+1, k) = \sum_{j=0}^{n} \binom{n}{j} S(j, k-1)$$

15. Show that the generating function

$$S_k(t) = \sum_0^\infty S(n+k, k)t^n$$

satisfies the recurrence

$$(1 - kt)S_k(t) = S_{k-1}(t)$$

so that

$$(1-t)(1-2t)\cdots(1-kt)S_k(t) = 1$$

Hence $S_0(t) = 1$, $S_1(t) = 1/(1-t)$, and by partial-fraction expansion

$$S_2(t) = \frac{2}{1-2t} - \frac{1}{1-t}$$

$$S_3(t) = \frac{9}{2(1-3t)} - \frac{4}{1-2t} + \frac{1}{2(1-t)}$$

$$S_k(t) = \frac{k^k}{k!}\frac{1}{1-kt} - \frac{(k-1)^k}{(k-1)!}\frac{1}{1-(k-1)t} + \cdots$$
$$+ \frac{(-1)^j}{k!}\binom{k}{j}\frac{(k-j)^k}{1-(k-j)t} + \cdots$$

[The last expression is obtained more easily when equation (38) in expanded form, namely;

$$S(n, k) = \frac{1}{k!}\Sigma(-1)^j\binom{k}{j}(k-j)^n$$

is used in the first equation of this problem.]

Note that for large values of $k$ the first term is dominant so that

$$\frac{S(n, k)}{k^n} \sim \frac{1}{k!}$$

where $\sim$ is the symbol for asymptotic equivalence.

16. Lah numbers.* Define the numbers $L_{n,k}$ by the equation

$$(-x)_n = \Sigma L_{n,k}(x)_k$$

so that

$$(x)_n = \Sigma L_{n,k}(-x)_k$$

* I. Lah, Eine neue Art von Zahren, ihre Eigenschaften und Anwendung in der mathematischen Statistik, *Mitteilungsbl. Math. Statist.*, vol. 7 (1955), pp. 203–212.

and, with $\delta_{n,m}$ a Kronecker delta,

$$\Sigma L_{n,k} L_{k,m} = \delta_{n,m}$$

Hence each of the equations

$$a_n = \Sigma L_{n,k} b_k$$
$$b_n = \Sigma L_{n,k} a_k$$

implies the other. Note that the numbers $(-1)^n L_{n,k}$ are positive.

(*a*) Derive the recurrence

$$L_{n+1,k} = -(n+k)L_{n,k} - L_{n,k-1}$$

and verify the table

| $n \backslash k$ | 1 | 2 | 3 | 4 | 5 |
|---|---|---|---|---|---|
| 1 | −1 | | | | |
| 2 | 2 | 1 | | | |
| 3 | −6 | −6 | −1 | | |
| 4 | 24 | 36 | 12 | 1 | |
| 5 | −120 | −240 | −120 | −20 | −1 |

(*b*) Write

$$L_k(t) = \sum_{n=0} L_{n,k}\, t^n/n!$$

From the relation

$$\sum_{n=0}^{\infty} (-x)_n t^n/n! = \sum_{k=0} \binom{x}{k} \left(\frac{-t}{1+t}\right)^k$$

show that

$$L_k(t) = \frac{1}{k!}\left(-\frac{t}{1+t}\right)^k$$

and, hence, that

$$L_{n,k} = (-1)^n \frac{n!}{k!} \binom{n-1}{k-1}$$

Verify that this result satisfies the recurrence in (*a*) and the boundary conditions.

(*c*) Derive the relation

$$\sum_{k=0} x^k L_k(-t) = \exp \frac{xt}{1-t}$$

(This result implies a relation of these numbers to those of Laguerre polynomials.)

(*d*) Show that

$$L_{n,k} = \Sigma(-1)^j s_{nj} S_{jk}$$

with $s_{nj}$ and $S_{jk}$ Stirling numbers.

17. If $a_0 = a_1 = 1$ and

$$a_{n+1} = (n+1)a_n - \binom{n}{2} a_{n-2}, \qquad n > 1$$

show that the exponential generating function

$$A(t) = a_0 + a_1 t + a_2 t^2/2! + \cdots$$

has the differential equation

$$(1 - t)\frac{dA}{dt} = (1 - \tfrac{1}{2}t^2)A(t)$$

and

$$A(t) = (1 - t)^{-\frac{1}{2}} \exp(t/2 + t^2/4)$$

18. Show that the operator $\theta^n$ ($n$ iterations of $\theta = tD$) satisfies

$$\theta^n = \sum_{k=0}^{n} S(n, k) t^k D^k$$

$$t^n D^n = \sum_{k=0}^{n} s(n, k)\theta^k = (\theta)_n$$

with $S(n, k)$, $s(n, k)$ Stirling numbers.

19. (*a*) From the generating function for Bernoulli numbers in the even suffix notation, namely

$$\exp tb = t/(e^t - 1) = [\log(1 + e^t - 1)](e^t - 1)^{-1}$$

or otherwise, show that

$$b_n = \sum_{k=0}^{n} (-1)^k k! S(n, k)/(k + 1)$$

(*b*) Show that

$$(b)_n = \Sigma\, s(n, k) b_k = (-1)^n n!/(n + 1)$$

[*Note*: This relation may be derived directly from the Bernoulli number generating function above and the substitution $t = \log(1 + u)$.]

20. By solving the system of equations (12), show that the element $a'_n$ of an inverse sequence may be written as the determinant

$$a'_n = (-1)^n a_0^{-n-1} \begin{vmatrix} a_1 & a_2 & a_3 & \cdots & a_{n-2} & a_{n-1} & a_n \\ a_0 & a_1 & a_2 & \cdots & a_{n-3} & a_{n-2} & a_{n-1} \\ 0 & a_0 & a_1 & \cdots & a_{n-4} & a_{n-3} & a_{n-2} \\ \cdot & \cdot & \cdot & \cdot & \cdot & \cdot & \cdot \\ 0 & 0 & 0 & \cdots & a_0 & a_1 & a_2 \\ 0 & 0 & 0 & \cdots & 0 & a_0 & a_1 \end{vmatrix}$$

21. Write $y$ for the set of $n$ variables $(y_1, y_2, \cdots y_n)$ ($y$ is a vector with $n$ components), $z$ for a similar set, and $y + z$ for the set $(y_1 + z_1, y_2 + z_2, \cdots, y_n + z_n)$.

Show that

$$Y_n(y + z) = \sum_{k=0}^{n} \binom{n}{k} Y_{n-k}(y) Y_k(z)$$

$$= (Y(y) + Y(z))^n, \qquad Y^k(y) \equiv Y_k(y), \qquad Y^k(z) \equiv Y_k(z)$$

If $a$ is a positive integer and $ay = (ay_1, \cdots, ay_n)$, show that

$$Y_n(ay) = (\underbrace{Y(y) + \cdots + Y(y)}_{\leftarrow\ a \text{ terms}\ \rightarrow})^n$$

22. Take $f(x) = x^m$, $g(x) = \exp cx = c_0 + c_1 x + c_2 x^2/2! + \cdots$ so that

$$A(x) = f(g(x)) = [\exp cx]^m = \exp xC(m), \qquad C^n(m) = C_n(m)$$

Show that

$$C_n(m) = Y_n(fc_1, fc_2, \cdots, fc_n), \qquad f_j = (m)_j c_0^{m-j}$$

Verify the instances

$$\begin{aligned} C_0(m) &= c_0^m \\ C_1(m) &= mc_1 c_0^{m-1} \\ C_2(m) &= mc_2 c_0^{m-1} + (m)_2 c_1^2 c_0^{m-2} \\ C_3(m) &= mc_3 c_0^{m-1} + 3(m)_2 c_2 c_1 c_0^{m-2} + (m)_3 c_1^3 c_0^{m-3} \end{aligned}$$

23. Write $(n)_{k,a} = n(n-a)\cdots(n-a(k-1)) = a^k(n/a)_k$, with the latter the usual falling factorial. Show that

$$(1 + at)^{n/a} = \sum_{k=0} (n)_{k,a} t^k/k!$$

and hence that

$$(n)_{k,a} = Y_k(fc_1, fc_2, \cdots, fc_k), \qquad f_j = (n)_j$$

with $c_0 = c_1 = 1$ and

$$c_j = (1)_{j,a} = (1-a)(1-2a)\cdots(1-(j-1)a), \qquad j > 1$$

Verify by direct calculation the particular results

$$\begin{aligned} (n)_{1,a} &= n = nc_1 \\ (n)_{2,a} &= n(n-a) = nc_2 + (n)_2 c_1^2 = n(1-a) + (n)_2 \\ (n)_{3,a} &= n(n-a)(n-2a) \\ &= n(1-a)(1-2a) + 3(n)_2(1-a) + (n)_3 \end{aligned}$$

24. If $a$ and $a'$ are inverse Blissard variables, so that

$$(\exp at)(\exp a't) = 1, \qquad a^n \equiv a_n, \qquad (a')^n = a'_n$$

show that

$$a'_n = Y_n(fa_1, fa_2, \cdots, fa_n), \qquad f_j = (-1)^j j! a_0^{-j-1}$$

and verify the results in the text following (23).

25. Take $f(u) = e^u$, $g(t) = at^2$, and

$$A(t) = f(g(t)) = e^{at^2}$$

Show that

$$A_n = D^n A(t) = e^{at^2} Y_n(g_1, g_2)$$

with $g_1 = 2at$, $g_2 = 2a$, ($g_n = 0$, $n > 2$). Hence find

$$\begin{aligned} \sum_{n=0} (e^{-at^2} D^n e^{at^2}) u^n/n! &= \exp(g_1 u + g_2 u^2/2!) \\ &= \exp au(2t + u) \end{aligned}$$

If

$$P_n(a, t) = e^{-at^2} D^n e^{at^2}$$

then $P_n(-1, t) = H_n(-t)$, where $H_n(t)$ is a Hermite polynomial (in a notation defined by this relation).

26. (*a*) For $A(t) = f(e^t)$, show that

$$A_n = D^n A(t) = \sum_{k=0}^{n} S(n, k) e^{kt} f_k$$

where $f_k = D_u^k f(u)|_{u=e^t}$

(*b*) With $f(u) = u^2$, so that $A(t) = e^{2t}$, verify that

$$A_n = 2^n e^{2t} = 2(S(n, 1) + S(n, 2))e^{2t}$$

27. (*a*) Show that

$$\log [1/(1 - \alpha_1 x)(1 - \alpha_2 x) \cdots] = s_1 x + s_2 x^2/2 + \cdots + s_n x^n/n + \cdots$$

with $s_n = \alpha_1^n + \alpha_2^n + \cdots$, a *power sum* symmetric function.

(*b*) Define the *elementary* symmetric functions $a_n$, $n = 1, 2, \cdots$ (which have appeared already in Section 1.4) and the *homogeneous product sum* symmetric functions $h_n$, $n = 1, 2, \cdots$ by the equations

$$\begin{aligned}(1 - \alpha_1 x)(1 - \alpha_2 x) \cdots &= 1 - a_1 x + a_2 x^2 - a_3 x^3 + \cdots \\ &= (1 + h_1 x + h_2 x^2 + h_3 x^3 + \cdots)^{-1}\end{aligned}$$

so that, using the result above,

$$1 - a_1 x + a_2 x^2 - \cdots = \exp(-s_1 x - s_2 x^2/2 - \cdots - s_n x^n/n - \cdots)$$
$$1 + h_1 x + h_2 x^2 + \cdots = \exp(s_1 x + s_2 x^2/2 + \cdots + s_n x^n/n + \cdots)$$

Show that

$$\begin{aligned}(-1)^n n!\, a_n &= Y_n(-s_1, -s_2, -2s_3, \cdots, -(n-1)!\, s_n) \\ n!\, h_n &= Y_n(s_1, s_2, 2s_3, \cdots, (n-1)!\, s_n) \\ -(n-1)!\, s_n &= Y_n(-fa_1, f2!a_2, -f3!a_3, \cdots, f(-1)^n n!\, a_n) \\ (n-1)!\, s_n &= Y_n(fh_1, f2!\, h_2, f3!\, h_3, \cdots, fn!\, h_n)\end{aligned}$$

with $f^j \equiv f_j = (-1)^{j-1}(j-1)!$

Verify the instances

$$\begin{aligned} a_1 &= s_1 & s_1 &= a_1 \\ 2a_2 &= -s_2 + s_1^2 & s_2 &= -2a_2 + a_1^2 \\ 6a_3 &= 2s_3 - 3s_2 s_1 + s_1^3 & 2s_3 &= 6a_3 - 6a_2 a_1 + 2a_1^3 \\ h_1 &= s_1 & s_1 &= h_1 \\ 2h_2 &= s_2 + s_1^2 & s_2 &= 2h_2 - h_1^2 \\ 6h_3 &= 2s_3 + 3s_2 s_1 + s_1^3 & 2s_3 &= 6h_3 - 6h_2 h_1 + 2h_1^3 \end{aligned}$$

28. (*a*) With $A_n(a) = A_n(a; g_1, g_2, \cdots, g_n)$ as in equation (43), and $D = d/dt$, show that

$$\begin{aligned}A_{n+1}(a) &= (ag_1 + D)A_n(a) \\ &= \left(ag_1 + \sum_1^n g_{s+1} \frac{d}{dg_s}\right) A_n(a)\end{aligned}$$

(*b*) Write

$$A_n(a) = \sum_{k=1}^{n} a^k A_{n,k}$$

Show that

$$A_{n+1,k} = g_1 A_{n,k-1} + DA_{n,k}$$

so that ($A_{n,0} = 0$), e.g.,

$$A_{n+1,1} = DA_{n1} = g_{n+1}$$
$$A_{n+1,2} = g_1 g_n + DA_{n,2}$$

From the last, derive in sequence

$$A_{2,2} = g_1^2 \qquad A_{4,2} = 4g_3 g_1 + 3g_2^2$$
$$A_{3,2} = 3g_2 g_1 \qquad A_{5,2} = 5g_4 g_1 + 10g_3 g_2$$

and, finally,

$$A_{n,2} = \sum_{j=0}^{n-2} \binom{n-1}{j} g_{j+1} g_{n-1-j}$$

TABLE 1

STIRLING NUMBERS OF THE FIRST KIND, $s(n, k)$

| $n\backslash k$ | 1 | 2 | 3 | 4 | 5 | 6 | 7 | 8 | 9 | 10 |
|---|---|---|---|---|---|---|---|---|---|---|
| 1 | 1 | | | | | | | | | |
| 2 | −1 | 1 | | | | | | | | |
| 3 | 2 | −3 | 1 | | | | | | | |
| 4 | −6 | 11 | −6 | 1 | | | | | | |
| 5 | 24 | −50 | 35 | −10 | 1 | | | | | |
| 6 | −120 | 274 | −225 | 85 | −15 | 1 | | | | |
| 7 | 720 | −1764 | 1624 | −735 | 175 | −21 | 1 | | | |
| 8 | −5040 | 13068 | −13132 | 6769 | −1960 | 322 | −28 | 1 | | |
| 9 | 40320 | −109584 | 118124 | −67284 | 22449 | −4536 | 546 | −36 | 1 | |
| 10 | −362880 | 1026576 | −1172700 | 723680 | −269325 | 63273 | −9450 | 870 | −45 | 1 |

TABLE 2

STIRLING NUMBERS OF THE SECOND KIND, $S(n, k)$

| $n\backslash k$ | 1 | 2 | 3 | 4 | 5 | 6 | 7 | 8 | 9 | 10 |
|---|---|---|---|---|---|---|---|---|---|---|
| 1 | 1 | | | | | | | | | |
| 2 | 1 | 1 | | | | | | | | |
| 3 | 1 | 3 | 1 | | | | | | | |
| 4 | 1 | 7 | 6 | 1 | | | | | | |
| 5 | 1 | 15 | 25 | 10 | 1 | | | | | |
| 6 | 1 | 31 | 90 | 65 | 15 | 1 | | | | |
| 7 | 1 | 63 | 301 | 350 | 140 | 21 | 1 | | | |
| 8 | 1 | 127 | 966 | 1701 | 1050 | 266 | 28 | 1 | | |
| 9 | 1 | 255 | 3025 | 7770 | 6951 | 2646 | 462 | 36 | 1 | |
| 10 | 1 | 511 | 9330 | 34105 | 42525 | 22827 | 5880 | 750 | 45 | 1 |

TABLE 3

BELL POLYNOMIALS $Y_n(fg_1, fg_2 \cdots fg_n)$

$$Y_1 = f_1 g_1$$

$$Y_2 = f_1 g_2 + f_2 g_1^2$$

$$Y_3 = f_1 g_3 + f_2(3g_2 g_1) + f_3 g_1^3$$

$$Y_4 = f_1 g_4 + f_2(4g_3 g_1 + 3g_2^2) + f_3(6g_2 g_1^2) + f_4 g_1^4$$

$$\begin{aligned} Y_5 = {} & f_1 g_5 + f_2(5g_4 g_1 + 10g_3 g_2) + f_3(10g_3 g_1^2 + 15g_2^2 g_1) \\ & + f_4(10g_2 g_1^3) + f_5 g_1^5 \end{aligned}$$

$$\begin{aligned} Y_6 = {} & f_1 g_6 + f_2(6g_5 g_1 + 15g_4 g_2 + 10g_3^2) \\ & + f_3(15g_4 g_1^2 + 60g_3 g_2 g_1 + 15g_2^3) \\ & + f_4(20g_3 g_1^3 + 45g_2^2 g_1^2) + f_5(15g_2 g_1^4) + f_6 g_1^6 \end{aligned}$$

$$\begin{aligned} Y_7 = {} & f_1 g_7 + f_2(7g_6 g_1 + 21g_5 g_2 + 35g_4 g_3) \\ & + f_3(21g_5 g_1^2 + 105g_4 g_2 g_1 + 70g_3^2 g_1 + 105g_3 g_2^2) \\ & + f_4(35g_4 g_1^3 + 210g_3 g_2 g_1^2 + 105g_2^3 g_1) \\ & + f_5(35g_3 g_1^4 + 105g_2^2 g_1^3) + f_6(21g_2 g_1^5) + f_7 g_1^7 \end{aligned}$$

$$\begin{aligned} Y_8 = {} & f_1 g_8 + f_2(8g_7 g_1 + 28g_6 g_2 + 56g_5 g_3 + 35g_4^2) \\ & + f_3(28g_6 g_1^2 + 168g_5 g_2 g_1 + 280g_4 g_3 g_1 + 210g_4 g_2^2 + 280g_3^2 g_2) \\ & + f_4(56g_5 g_1^3 + 420g_4 g_2 g_1^2 + 280g_3^2 g_1^2 + 840g_3 g_2^2 g_1 + 105g_2^4) \\ & + f_5(70g_4 g_1^4 + 560g_3 g_2 g_1^3 + 420g_2^3 g_1^2) \\ & + f_6(56g_3 g_1^5 + 210g_2^2 g_1^4) + f_7(28g_2 g_1^6) + f_8 g_1^8 \end{aligned}$$

CHAPTER 3

# The Principle of Inclusion and Exclusion

## 1. INTRODUCTION

This chapter is devoted to an important combinatorial tool, the principle of inclusion and exclusion, also known variously as the symbolic method, principle of cross classification, sieve method (the significance of these terms will become apparent later). The logical identity on which it rests is very old; Dickson's *History of the Theory of Numbers* (vol. I, p. 119) mentions its appearance in a work by Daniel da Silva in 1854, but Montmort's solution in 1713 of a famous problem, known generally by its French name, "le problème des rencontres" (the number of permutations of $n$ elements such that no element is in its own position), effectively uses it and it may have been known to the Bernoullis.

The present concern is with general formulations, some of which are foreshadowed by the discussion of factorial moments in Chapter 2, which will clear the ground for the many applications which are to follow.

## 2. A LOGICAL IDENTITY

Suppose there are $N$ objects, and $N(a)$ (using functional notation) have the property $a$; then, if $a'$ denotes the absence of property $a$,

$$N(a') = N - N(a)$$

for each object either has or has not property $a$. If two properties $a$ and $b$ are in question, the number without both is given by

$$N(a'b') = N - N(a) - N(b) + N(ab)$$

for, in subtracting $N(a)$ and $N(b)$ from the total, $N(ab)$ has been subtracted twice and must be restored. This justifies the term "inclusion and exclusion"; the process is one of including everything, excluding those not required, including those wrongly excluded, and so on, alternately including and excluding.

The general result may be stated as:

**Theorem 1.** *If of $N$ objects, $N(a)$ have property $a$, $N(b)$ property $b, \ldots, N(ab)$ both $a$ and $b, \ldots, N(abc)$ $a$, $b$, and $c$, and so on, the number $N(a'b'c' \ldots)$ with none of these properties is given by*

$$\begin{aligned} N(a'b'c' \cdots) = N - N(a) - N(b) - \cdots \\ + N(ab) + N(ac) + \cdots \\ - N(abc) - \cdots \\ + \cdots \end{aligned} \tag{1}$$

The proof of (1) by mathematical induction is simple once it is noted that the formula $N(a') = N - N(a)$ can be applied to any collection of the objects which is suitably defined.

It is convenient to relabel the properties $a_1, a_2, \cdots, a_n$. Then first, by the remark just above, $N(a_1'a_2' \cdots a_{n-1}'a_n') = N(a_1'a_2' \cdots a_{n-1}') - N(a_1'a_2' \cdots a_{n-1}'a_n)$. Next, suppose that (1) is true for the $n - 1$ properties $a_1$ to $a_{n-1}$ so that

$$\begin{aligned} N(a_1'a_2' \cdots a_{n-1}') = N - N(a_1) - N(a_2) - \cdots - N(a_{n-1}) \\ + N(a_1a_2) + \cdots + N(a_{n-2}a_{n-1}) \\ + \cdots + (-1)^{n-1}N(a_1a_2 \cdots a_{n-1}) \end{aligned}$$

Applied to the collection $N(a_n)$ this gives

$$\begin{aligned} N(a_1'a_2' \cdots a_{n-1}'a_n) = N(a_n) - N(a_1a_n) - N(a_2a_n) - N(a_{n-1}a_n) \\ + N(a_1a_2a_n) + \cdots \\ + (-1)^{n-1}N(a_1a_2 \cdots a_{n-1}a_n) \end{aligned}$$

Subtraction of these two equations, by the first result, gives equation (1) for $n$ properties $a_1$ to $a_n$. Hence, if the theorem is true for $n - 1$ properties, it is true for $n$, and, since it is true for $n = 1$, it is true for any $n$, by mathematical induction.

For ease in using the theorem to its full advantage, the following remarks are important.

First, it seems most easily remembered in the following symbolic form

$$N(a'b'c' \cdots) = N[(1 - a)(1 - b)(1 - c) \cdots] \tag{1a}$$

The meaning of the symbolic expression on the right is that the product in brackets is first multiplied out and then terms are separated as in

$$\begin{aligned} N[(1 - a)(1 - b)] &= N(1 - a - b + ab) \\ &= N(1) - N(a) - N(b) + N(ab) \end{aligned}$$

with $N(1) = N$. More formally, the following conventions are used for interpretation: $N(a + b) = N(a) + N(b)$, $N(-a) = -N(a)$, $N(1) = N$. In logic, $a'$ is called the complement of $a$ and is defined by $a' + a = 1$, where 1 stands for the totality in question, which makes equation ($1a$) intuitively natural.

Next, as mentioned in the proof, the totality of objects to which the theorem is applied need not be the full collection of $N$; instead, any subset suitably defined by the properties may be used. For example, in addition to those already mentioned

$$N(a_1a_2'a_3a_4') = N[a_1(1 - a_2)a_3(1 - a_4)]$$
$$= N(a_1a_3) - N(a_1a_2a_3) - N(a_1a_3a_4) + N(a_1a_2a_3a_4)$$

In general, if the collection of properties $a_1, a_2 \cdots a_n$ is relabeled in any way $b_1b_2 \cdots b_kc_1c_2 \cdots c_{n-k}$, then

$$N(b_1b_2 \cdots b_kc_1'c_2' \cdots c_{n-k}') = N[b_1b_2 \cdots b_k(1 - c_1) \cdots (1 - c_{n-k})] \quad (1b)$$

The formula is chiefly interesting when one or more of the properties are compatible, that is, when some of the terms $N(a_ja_k)$, $N(a_ja_ka_l)$, and so on, are not zero. When all properties are mutually exclusive, the contrary case, equation (1) appears in the degenerate form

$$N(a_1'a_2' \cdots a_n') = N - N(a_1) - N(a_2) - \cdots N(a_n)$$

a relatively uninteresting result.

In probability theory, (1) goes under the name of Poincaré's theorem and its symbolic form is given as

$$P(a_1'a_2' \cdots a_n') = [1 - P(a_1)][1 - P(a_2)] \cdots [1 - P(a_n)] \quad (1c)$$

The symbolic interpretation of the right-hand side is like that of ($1a$): after multiplication, $P(a_i)P(a_j) \cdots$ is replaced by $P(a_ia_j \cdots)$. The complete equation is re-interpreted in terms of probability; the properties $a_1, a_2 \cdots$ describe "events" instead of objects, $P(a_1a_2)$ is the probability of occurrence of the joint event described by $a_1$ and $a_2$, and so on. For independent events, ($1c$) is true without symbolic interpretation and, hence, may be regarded as a generalization in a natural way of the independent events formula to events which are dependent and compatible (see Frechet, **1** and **2**).

## 3. SYMBOLIC DEVELOPMENT

The identity (1) may be given a considerable symbolic development (which justifies the term "symbolic method"). For this purpose, label the properties $a_1$ to $a_n$ as before and define $S_0 = 1$,

$$S_1 = N^{-1}\Sigma N(a_i)$$
$$S_2 = N^{-1}\Sigma N(a_ia_j), \qquad i \neq j$$
$$S_3 = N^{-1}\Sigma N(a_ia_ja_k), \qquad i, j, k \text{ unequal}$$

and so on. The sum in $S_1$ is over all $n$ properties, in $S_2$ over all pairs, in $S_3$ over all triples, and so on. Then, if $S_{n+k} = 0$, $k > 0$, (1) may be written

$$N^{-1}N(a'_1a'_2 \cdots a'_n) = 1 - S_1 + S_2 - S_3 + \cdots + (-1)^n S_n$$
$$= (1 + S)^{-1}, \qquad S^k \equiv S_k \tag{2}$$

This is a probability form, dealing with relative rather than absolute numbers; the term on the left is the relative number with none of the properties and may be replaced by $p(0)$. In the same notation, $p(k)$ is the relative number of objects with exactly $k$ of the given properties, the $k$ properties in question being any of the $\binom{n}{k}$ selections which are compatible. Thus

$$p(k) = \Sigma N^{-1}N(a_1a_2 \cdots a_k a'_{k+1} \cdots a'_n)$$

if the sum is over all possible selections of the $n$ properties $k$ at a time; note that, to avoid elaboration of notation, the term indicated in the sum is the first of such choices and not the general one.

Then by (1*b*) and with the same convention as in equation (2),

$$p(k) = S_k - (k+1)S_{k+1} + \binom{k+2}{2} S_{k+2} - \cdots + (-1)^{n-k}\binom{n}{k} S_n$$
$$= S^k(1 + S)^{-k-1}, \qquad S^k \equiv S_k \tag{3}$$

Binomial coefficients appear because, for example, a term $N(a_1a_2 \cdots a_{k+j})$ appears as many times as there are $k$-combinations of $k + j$ things. Equation (2) is the special case $k = 0$ of equation (3).

Equation (3), apart from notation, is identical with equation (2.32) expressing probabilities in terms of their binomial moments, namely,

$$p_k = B_k - (k+1)B_{k+1} + \binom{k+2}{2} B_{k+2} - \cdots$$

Hence the sums $S_k$ are the binomial moments of the probabilities $p(k)$, that is,

$$S_k = \Sigma \binom{j}{k} p(j) \tag{4}$$
$$= p(k) + (k+1)p(k+1) + \binom{k+2}{2} p(k+2) + \cdots + \binom{n}{k} p(n)$$

The generating function of $p(k)$, namely,

$$P(t) = p(0) + p(1)t + p(2)t^2 + \cdots$$

by the second of equations (2.30) is given by

$$P(t) = \exp(m)(t-1) = \Sigma B_k(t-1)^k = \Sigma S_k(t-1)^k \tag{5}$$
$$= [1 - S(t-1)]^{-1}, \qquad S^k \equiv S_k$$

This is verified by the symbolic form of (3), since

$$\begin{aligned} P(t) &= \sum_{k=0} S^k(1+S)^{-k-1}t^k \\ &= (1+S)^{-1} \sum_{k=0} [St/(1+S)]^k \\ &= \frac{1}{1+S-St}, \qquad S^k \equiv S_k \end{aligned}$$

It may also be verified by direct calculation with the finite sums appearing in (3), $k = 0, 1, 2, \cdots$.

It is also worth noting that the absolute rather than the relative numbers have the generating function

$$\begin{aligned} N(t) = NP(t) &= N[1 + S(t-1)]^{-1}, \qquad S^k \equiv S_k \\ &= \sum_{k=0}^{n} NS_k(t-1)^k \end{aligned} \tag{5a}$$

($N(t)$ as a generating function, of course, is not to be confused with $N(a_i)$, the number of objects with property $a_i$).

The variable for the cumulation of probabilities $p(k)$ may be taken as any of

$$\begin{aligned} u(k) &= p(0) + p(1) + \cdots + p(k) \\ v(k) &= p(k+1) + p(k+2) + \cdots + p(n) \\ r(k) &= p(k) + p(k+1) + \cdots = p(k) + v(k) \end{aligned}$$

Note that $u(k) + v(k) = 1$. The corresponding generating functions are

$$U(t) = (1-t)^{-1}P(t) = -\Sigma S_k(t-1)^{k-1} \tag{6}$$

$$V(t) = (1-t)^{-1}[1 - P(t)] = \Sigma(S_{k+1}(t-1)^k \tag{7}$$

$$R(t) = (1-t)^{-1}[1 - tP(t)] = \Sigma(S_{k+1} + S_k)(t-1)^k \tag{8}$$

and the symbolic expressions are

$$u(k) = 1 - S^{k+1}(1+S)^{-k-1}, \qquad S^n \equiv S_n \tag{9}$$

$$v(k) = S^{k+1}(1+S)^{-k-1}, \qquad S^n \equiv S_n \tag{10}$$

$$r(k) = S^k(1+S)^{-k}, \qquad S^n \equiv S_n \tag{11}$$

More generally, it may be supposed that a gain $g_k$ is realized by the appearance of an "event" $k$, that is, an object with exactly $k$ of the $n$

properties, in some game with two or more players; then the expected gain is a function of $g_1, g_2, \cdots, g_n$ defined by

$$G(g_0, g_1, \cdots, g_n) = g_0p(0) + g_1p(1) + \cdots + g_np(n) \qquad (12)$$

Note that

$$G(1, t, \cdots, t^n) = P(t)$$

and that $G$ is a generating function for the sequence $p(k)$ in the general sense of Chapter 2. Following Touchard, **3**, this may be given another form, which may be stated as

**Theorem 2.** *If*

$$G(g_1, g_2, \cdots, g_n) = g_0p(0) + g_1p(1) + \cdots + g_np(n)$$

*and*

$$S_k = \Sigma \binom{j}{k} p(j)$$

$$\Delta g_0 = g_1 - g_0$$

$$\Delta^k g_0 = \Delta(\Delta^{k-1}g_0) = (g - 1)^k, \qquad g^j \equiv g_j$$

*then*

$$G(g_1, g_2, \cdots, g_n) = g_0S_0 + (\Delta g_0)S_1 + \cdots + (\Delta^n g_0)S_n$$

The theorem is proved at once by noting that (by definition) $S_k$ is a binomial moment; hence, by equation (2.32)

$$p(j) = \Sigma(-1)^k \binom{k+j}{k} S_{j+k} = S^k(1 + S)^{-k-1}, \qquad S^n \equiv S_n$$

and

$$\begin{aligned} G(g_0, g_1, \cdots, g_n) &= \Sigma g_jp(j) \\ &= \Sigma[gS/(1 + S)]^k(1 + S)^{-1}, \qquad g^j \equiv g_j, \qquad S^j \equiv S_j \\ &= [1 - S(g - 1)]^{-1} \\ &= \Sigma S^k(g - 1)^k = \Sigma S_k\, \Delta^k g_0 \end{aligned}$$

It may be noted that for $g_k = t^k$, $\Delta^k g_0 = (t - 1)^k$ and

$$G(1, t, \cdots, t^n) = \Sigma S_k(t - 1)^k$$

which is $P(t)$ in the form of equation (5).

Another verification is obtained by setting, arbitrarily,

$$G(g_0, g_1, \cdots, g_n) = S_k$$

whence, by the conclusion of the theorem, with $\delta_{jk}$ a Kronecker delta,

$$\Delta^j g_0 = \delta_{jk}$$

and

$$\begin{aligned}\exp t(g-1) &= \Sigma t^j (g-1)^j/j! \\ &= \Sigma t^j \, \Delta^j g_0/j! \\ &= t^k/k!\end{aligned}$$

But this is equivalent to

$$\exp tg = e^t t^k/k! = \Sigma \binom{j}{k} t^j/j!$$

or

$$g_j = \binom{j}{k}$$

a verification.

Another form of the theorem is obtained by replacing $p(j)$ by $p(j)t^j$, so that, if

$$P(t) = p(0) + p(1)t + p(2)t^2 + \cdots$$

then

$$S(k) = \Sigma \binom{j}{k} p(j)t^j = t^k P^{(k)}(t)/k!$$

with $P^{(k)}(t)$ the $k$th derivative. Hence

**Theorem 2a.** *If*

$$G(g_0, g_1, \cdots) = g_0 p(0) + g_1 p(1)t + \cdots + g_j p(j)t^j + \cdots$$

*and*

$$P(t) = p(0) + p(1)t + \cdots$$

*then*

$$\begin{aligned}G(g_0, g_1, \cdots) = g_0 P(t) &+ \Delta g_0 P'(t)t + \cdots \\ &+ \Delta^k g_0 P^{(k)}(t)t^k/k! + \cdots\end{aligned}$$

It may be remarked that Euler's transformation of series is a special case of Theorem 2*a*.

## 4. RANK

The enumerations made above are of objects with a given number of properties, without any implication of rank or order in these properties. In this section the properties are ordered and the objects enumerated according to the rank of the first property present; that is, the objects of rank $k$ are enumerated by $N(a_1' \cdots a_{k-1}' a_k)$. To complete the enumeration, the objects enumerated by $N(a_1' \cdots a_n')$ are said, by convention, to be of rank $n + 1$.

Take $R(j)$ for the number of objects of rank $j$, $j = 1, 2, \cdots, n + 1$, and $r(j) = N^{-1}R(j)$ for the relative number; the number of properties $n$

is here carried silently but, of course, may be indicated explicitly when needed.

For concreteness, note that for $n = 2$

$$R(1) = N(a_1)$$
$$R(2) = N(a_1'a_2) = N(a_2) - N(a_1a_2)$$
$$R(3) = N(a_1'a_2') = N - N(a_1) - N(a_2) + N(a_1a_2)$$

In general, by equation (1*b*),

$$\begin{aligned} R(j) &= N(a_1'a_2' \cdots a_{j-1}'a_j) \qquad (13) \\ &= N[(1-a_1)(1-a_2)\cdots(1-a_{j-1})a_j] \\ &= N(a_j) - N(a_1a_j) - \cdots - N(a_{j-1}a_j) + \cdots \\ &\qquad + (-1)^{j-1}N(a_1a_2\cdots a_j) \end{aligned}$$

If the probability interpretation is followed, and $P(a_1a_2 \cdots a_j)$ is the probability that a success is achieved for each of trials 1 to $j$, then $r(j) = P(a_1'a_2' \cdots a_{j-1}'a_j)$ is the probability that the first success is achieved on the $j$th trial. For independent trials, $P(a_1a_2 \cdots a_j) = p^j$, if $p$ is the probability of success on any trial, and $P(a_1'a_2' \cdots a_{j-1}'a_j) = (1-p)^{j-1}p$, a simple and well-known result.

Some simplification appears if symmetry is present, that is, if

$$N^{-1}N(b_1b_2 \cdots b_k) = s_k$$

where $b_1b_2 \cdots b_k$ is *any* selection of $k$ of the $n$ properties $a_1$ to $a_n$; then

$$\begin{aligned} r(j) &= s_1 - (j-1)s_2 + \binom{j-1}{2}s_3 - \cdots + (-1)^{j-1}s_j, \qquad 0 < j < n+1 \\ &= s(1-s)^{j-1}, \qquad s^k \equiv s_k \qquad (14) \end{aligned}$$

and

$$r(n+1) = (1-s)^n, \qquad s^k \equiv s_k$$

These expressions may be used to form generating functions, just as in Section 3, as will appear in the next section and in the problems.

## 5. THE "PROBLEME DES RENCONTRES"

The material given above may be illustrated by the "problème des rencontres". As already mentioned, in its simplest form this asks for the number of permutations of $n$ distinct elements, which we take as the numbers 1 to $n$, such that element $k$ is not in the $k$th position, $k = 1$ to $n$. More generally, it is the problem of enumerating permutations of $n$

distinct elements, according to the number of elements in "their own positions"

Some practical settings of the problem may be noted. In its original form (Montmort), a permutation was formed by the drawing of numbered balls from an urn after mixing, as in a lottery. A later version is of letters and envelopes with random pairing; this is the same as matching two decks of cards, each consisting of $n$ distinct cards. Or suppose the pages of a manuscript are scattered, by the wind, a small child, or other demonic activity, and reassembled in a hurry.

Thinking of card matching, the enumeration is of number of elements displaced (in one deck relative to the other) and is said to be of displacements (which justifies the notation $D_{n,k}$, $D_n(t)$, and so forth, followed below). The complementary point of view is in terms of elements not displaced, which are called "hits" or coincidences, as when the subject calls the right card in telepathy experiments.

The totality of objects to be considered is the number of permutations of $n$ distinct elements, so $N = n!$. These are distinguished according to the $n$ properties $a_1$ to $a_n$, where $a_k$ means that element $k$ is in the $k$th position. Hence the problem is symmetric, since the selection of a set of $j$ properties determines a number independent of the particular set selected; thus, if $b_1, b_2, \cdots, b_j$ is any selection of $j$ of the $n$ properties,

$$N(b_1 b_2 \cdots b_j) = (n - j)!$$

since fixing the position of $j$ elements leaves $(n - j)!$ permutations for the remaining elements. Then the relative sum, $S_j$, over all such selections, as defined in Section 3, turns out to be

$$S_j = \binom{n}{j} (n - j)!/n! = 1/j! \tag{15}$$

Using the results of Section 3, the problem is now completely solved. Thus, by equations (2) and (3), indicating dependence on $n$ by a subscript,

$$\begin{aligned} p_n(0) &= 1 - S_1 + S_2 - S_3 + \cdots + (-1)^n S_n \\ &= 1 - 1 + 1/2! - 1/3! + \cdots + (-1)^n/n! \end{aligned} \tag{16}$$

$$\begin{aligned} p_n(k) &= S_k - (k + 1)S_{k+1} + \binom{k+2}{2} S_{k+2} - \cdots + (-1)^{n-k} \binom{n}{k} S_n \\ &= [1 - 1 + 1/2! - \cdots + (-1)^{n-k}/(n - k)!]/k! \\ &= p_{n-k}(0)/k! \end{aligned} \tag{17}$$

Thus all probabilities are determined by $p_n(0)$, which by (16) is the truncated exponential series $e^{-1}$. It may be noted that the mean of the distribution is $S_1 = 1$, and that, for large $n$, the distribution is approximately Poisson with mean $1 (p_n(k) \sim e^{-1}/k!)$.

For small values of $n$, it is convenient to deal with the absolute numbers $D_{n,k} = n!p_n(k)$ mentioned above. Note that $D_{n,k}$ enumerates the number of permutations of $n$ elements with exactly $k$ *not* displaced. Then by (17)

$$D_{n,k} = \binom{n}{k} D_{n-k,0} = \binom{n}{k} D_{n-k} \tag{17a}$$

Writing $D_{n,0} \equiv D_n$, the generating function

$$D_n(t) = \sum_0^n D_{n,k} t^k$$

may be written as the Appell polynomial

$$D_n(t) = (D + t)^n, \qquad D^n \equiv D_n \tag{18}$$

This also follows from (5), since

$$D_n(t) = n!P_n(t) = n! \sum_0^n (t-1)^k/k!$$

and, with a prime denoting a derivative,

$$D_n'(t) = nD_{n-1}(t) \tag{19}$$

the defining relation for Appell polynomials. Note also that

$$D_n(t) = nD_{n-1}(t) + (t-1)^n \tag{20}$$

from which follows at once, by setting $t = 0$, the simplest recurrence for the numbers $D_n \equiv D_n(0)$, namely,

$$D_n = nD_{n-1} + (-1)^n \tag{21}$$

Because of the similarity of this to the recurrence for factorials, Whitworth and others call the numbers $D_n$ subfactorials. It may also be noticed that, by (16) [or by iteration of (21)]

$$D_n = \sum_0^n \binom{n}{k} (-1)^k (n-k)! = (E-1)^n 0! = \Delta^n 0! \tag{22}$$

with $E$ and $\Delta$ finite difference operators. This may be contrasted with $n! = E^n 0!$ in justification of the subfactorial term.

The numbers $D_{n,k}$ are given in Table 1 of this chapter for $n = 0(1)10$, $k = 0(1)n$.

As expected in so simple a problem, all this can be proved in a number of ways by direct methods. First, (17$a$) is found by noticing that $k$ positions of coincidence may be chosen in $\binom{n}{k}$ ways, and for each such choice the number of permutations of the remaining $n - k$ elements without coincidence is $D_{n-k,0}$, by definition. Equation (18) follows just as before. Next, since

$$n! = \sum_{0}^{n} D_{n,k} = \Sigma \binom{n}{k} D_{n-k} = (D + 1)^n, \qquad D^k \equiv D_k$$

it follows that

$$\exp u(D + 1) = (1 - u)^{-1} \tag{23}$$

or

$$(1 - u) \exp uD = e^{-u}$$

which gives (21) by equating coefficients of $u^n/n!$.

Similarly the equation

$$\exp uD = e^{-u}(1 - u)^{-1}$$

gives (22) by direct expansion.

Euler's determination of the number of permutations of $n$ elements with all elements displaced, that is, of numbers $D_n$, is as follows. The first position is open to all elements save 1, hence to $n - 1$ elements. Suppose the element in the first position is $k$ ($k \neq 1$). Then, the permutations of the remaining elements are of two classes according as 1 is or is not in the $k$th position. If 1 is in the $k$th position, the number of permutations is that of $n - 2$ elements with all elements displaced, that is, $D_{n-2}$. If 1 is not in from the $k$th position, the permutations permitted are those of elements $1, 2, \cdots, (k - 1), (k + 1), \cdots, n$ into positions 2 to $n$ with 1 not in the $k$th position and every other element not in its own position. But this is the same as the permutations of $n - 1$ elements 2 to $n$ with every element displaced, enumerated by $D_{n-1}$. Hence

$$D_n = (n - 1)(D_{n-1} + D_{n-2}) \tag{24}$$

which in the form

$$D_n - (n)D_{n-1} = -(D_{n-1} - (n - 1)D_{n-2}) \tag{24a}$$

leads by iteration to (21) since $D_2 = 1$. Note that $n!$ also satisfies recurrence (24), another formal similarity of $D_n$ and $n!$.

Turning to the enumeration by rank, note first that, as remarked above, the properties are symmetric, so that the relative number of permutations with a given set of $k$ elements in their own positions, the $s_k$ of equation (14), is given by

$$s_k = (n-k)!/n!$$

Hence, using equation (14)

$$\begin{aligned} r(j) &= s_1 - (j-1)s_2 + \binom{j-1}{2} s_3 - \cdots + (-1)^{j-1}s_j, \qquad 0 < j < n+1 \\ &= \frac{1}{n!}\sum_{0}^{j-1}(-1)^k \binom{j-1}{k}(n-1-k)! \end{aligned} \tag{25}$$

and

$$\begin{aligned} r(n+1) &= (1-s)^n = \sum_{0}^{n}\binom{n}{k}(-1)^k s_k \\ &= 1 - 1 + \frac{1}{2!} - \frac{1}{3!} + \cdots + \frac{(-1)^n}{n!} \\ &= D_n/n! \end{aligned} \tag{26}$$

Equation (25) may be written more concisely in a finite difference form; thus, since $(n-1-k)! = E^{j-1-k}(n-j)!$

$$r(j) = \Delta^{j-1}(n-j)!/n!, \qquad 0 < j < n+1 \tag{25a}$$

The absolute numbers are then given by

$$R_n(j) = \Delta^{j-1}(n-j)!, \qquad j = 1, 2, \cdots, n \tag{27}$$

$$R_n(n+1) = D_n = \Delta^n 0! \tag{28}$$

Note that $R_n(1) = (n-1)!$. A recurrence relation follows from (27) as follows:

$$\begin{aligned} R_n(j) &= (E-1)^{j-1} E^{n-j}0! \\ &= (E-1)^{j-2} E^{n+1-j}0! - (E-1)^{j-2} E^{n-j}0! \\ &= R_n(j-1) - R_{n-1}(j-1), \qquad 1 < j < n+1 \end{aligned} \tag{29}$$

Changing notation: $R_n(j) = R_{nj}$ to avoid confusion with the ordinary generating-function notation:

$$R_n(t) = R_{n1}t + R_{n2}t^2 + \cdots + R_{n,n+1}t^{n+1}$$

and using (29) and $R_{nn} = D_{n-1}$, it is found that

$$(1-t)R_n(t) = t[(n-1)! - R_{n-1}(t)] + D_n t^{n+1}(1-t) \tag{30}$$

For $t = 1$, this gives $R_{n-1}(1) = (n-1)!$, a verification, and differentiation shows that, denoting a derivative by a prime,

$$R'_{n-1}(1) = n! - D_n$$

Hence the expected rank, for permutations of $n$, is

$$E[r(j)] = R'_n(1)/n! \qquad (31)$$
$$= n + 1 - D_{n+1}/n!$$
$$\sim (n + 1)(1 - e^{-1})$$

a result given in Frechet, **2**.

## REFERENCES

**1.** Frechet, Maurice, *Les probabilités associées à un system d'événements compatibles et dépendants.* Part I: *Événements en nombre fini fixe*, Paris, 1940.
**2.** ———, Part II. *Cas particuliers et applications*, Paris, 1943.
**3.** Touchard, Jacques, Remarques sur les probabilités totales et sur le problème des rencontres, *Ann. Soc. Sci., Bruxelles*, vol. A53 (1933), pp. 126–134.
**4.** Whitney, Hassler, A logical expansion in mathematics, *Bull. Amer. Math. Soc.*, vol. 38 (1932), pp. 572–579.

## PROBLEMS

1. Show that the number of integers equal to or less than $N$ which are not divisible by the primes 2, 3, 5, 7 is

$$N - \Sigma[N/i] + \Sigma[N/ij] - \Sigma[N/ijk] + \Sigma[N/ijkl]$$

where $[N/i]$ is the greatest integer not greater than $N/i$ and the sums are over all four given primes, all unlike pairs of these, and so on. If $N = 210n$, show that this number is $48n$.

2. Take $p_1, p_2, \cdots, p_r$ for the primes not greater than $\sqrt{N}$. Show that the number of primes greater than $p_r$ and not greater than $N$ is

$$N - 1 - S_1 + S_2 - \cdots + (-1)^r S_r$$

with

$$S_k = \Sigma\,[N/p_1^{a_1} p_2^{a_2} \cdots p_r^{a_r}]$$

the sum being over all $\binom{r}{k}$ ways of setting $k$ of $a_1, a_2, \cdots, a_r$ equal to one and the rest equal to zero.

3. Show that $\phi(n)$, the number of positive integers less than $n$ and prime to $n$ (Euler's totient function), is given by

$$\phi(n) = n\left(1 - \frac{1}{p_1}\right)\left(1 - \frac{1}{p_2}\right)\cdots$$

where $p_1, p_2, \cdots$ are the different prime factors of $n$. Verify the following table ($\phi(1) = 1$ is a convention):

| $n$ | 1 | 2 | 3 | 4 | 5 | 6 | 7 | 8 | 9 | 10 |
|---|---|---|---|---|---|---|---|---|---|---|
| $\phi(n)$ | 1 | 1 | 2 | 2 | 4 | 2 | 6 | 4 | 6 | 4 |

4. For independent events $a_1$ to $a_n$, each with probability $p$, it follows from (1$c$) that $P(a_1'a_2' \cdots a_n') = (1-p)^n$ and, hence, that $S_k = \binom{n}{k}p^k$. Show that

$$\begin{aligned} r(k) &= p(k) + p(k+1) + \cdots p(n) \\ &= S^k(1+S)^{-k}, \qquad S^j \equiv S_j \\ &= n\binom{n-1}{k-1}\sum_0^{n-k}(-1)^j\binom{n-k}{j}\frac{p^{k+j}}{k+j} \\ &= n\binom{n-1}{k-1}\int_0^p x^{k-1}(1-x)^{n-k}\,dx \end{aligned}$$

5. For independent events as in Problem 4, show that the rank variable $r_{n,j}$ is given by

$$r_{n,j} = pq^{j-1}, \qquad j < n+1$$
$$r_{n,n+1} = q^n$$

with $q = 1 - p$. Show that the generating function and mean are

$$r_n(t) = [pt + (1-t)(qt)^{n+1}](1-qt)^{-1}$$
$$\bar{r}_{n,j} = (1-q^{n+1})/p$$

6. ($a$) From the relations, equations (18) and (23),

$$D_n(t) = (D+t)^n, \qquad D^n \equiv D_n$$
$$\exp u(D+1) = (1-u)^{-1}, \qquad D^n \equiv D_n$$

show that

$$\exp uD(t) = (1-u)^{-1}\exp u(t-1), \qquad D^n(t) \equiv D_n(t)$$

($b$) Inverting the last relation show that

$$n! = \sum_0^n \binom{n}{k}(1-t)^k D_{n-k}(t)$$

and verify from this relation that $D_0(t) = 1$, $D_1(t) = t$, $D_2(t) = 1 + t^2$.

($c$) For $n = 3$ and 4, exhibit the permutations enumerated by $D_{n,k}$ for all values of $k$.

($d$) Using equation (20), or otherwise, derive the recurrence

$$D_{n+1}(t) = (n+t)D_n(t) + n(1-t)D_{n-1}(t)$$

7. For the rank numbers $R_n(j) \equiv R_{nj}$ of the "problème des rencontres" show that, with a prime denoting a derivative,

$$R_{nj} = D^{j-1}(D+1)^{n-j}, \qquad j < n+1, \qquad D^k \equiv D_k$$
$$R_{nj} = (n-j)R_{n-1,j} + (j-1)R_{n-1,j-1}, \qquad j < n$$
$$R_n(t) = nR_{n-1}(t) - t(1-t)[R'_{n-1}(t) + (-1)^n t^{n-1}]$$

8. If $(p_k)$ is a sequence of probabilities, and $(B_k)$ the sequence of its binomial moments, show that

$$p_1 + p_3 + \cdots + p_{2k+1} + \cdots = B_1 - 2B_2 + \cdots + (-2)^{k-1}B_k + \cdots$$
$$p_2 + p_3 + p_5 + p_7 \cdots = B_2 - 2B_3 + 2B_4 + B_5 - 11B_6 + 36B_7 - 92B_8 + \cdots$$

The first series in $p_k$ has all odd indexes; the second has all indexes prime numbers. (See Touchard, **3**.)

9. *Generalization of rencontres.*

(*a*) If permutations of $n$ distinct elements are considered in batches of $m$, so that there are $n!^m$ possible batches, show that the number of batches in which no position contains its own element in every permutation of the batch is $\Delta^n 0!^m$, that is,

$$\sum \binom{n}{k} (-1)^k (n-k)!^m \qquad \text{(Laplace)}$$

(*b*) With $p_k$ the number of batches with $k$ elements in their own positions in each permutation of the batch, show that

$$p_k = \binom{n}{k} \Delta^{n-k} 0!^m$$

$$P(t) = \Sigma p_k t^k = (\Delta + t)^n 0!^m$$

(*c*) With $D_n(m) = \Delta^n 0!^m$, show that

$$D_n(m) = n \sum_0^{m-1} \binom{n-1}{k} \Delta^k (n-k)^{m-1} D_{n-1-k}(m) + (-1)^n$$

and verify by the table:

| $n$ | 0 | 1 | 2 | 3 | 4 | 5 |
|---|---|---|---|---|---|---|
| $D_n(1)$ | 1 | 0 | 1 | 2 | 9 | 44 |
| $D_n(2)$ | 1 | 0 | 3 | 26 | 453 | 11844 |
| $D_n(3)$ | 1 | 0 | 7 | 194 | 13005 | 1660964 |

(*d*) Using the notation of (*c*) and the relations

$$n!^m = \sum \binom{n}{k} D_{n-k}(m)$$

$$(x+y)^p \equiv x^p + y^p, \qquad \text{mod } p$$

with $p$ a prime number, show that

$$D_{n+p}(m) \equiv -D_n(m), \qquad \text{mod } p$$

$$D_{n+2p}(m) \equiv D_n(m), \qquad \text{mod } p$$

(The residues, mod $p$, repeat with period $2p$.)

10. *Lottery.* $m$ balls at a time are drawn from an urn containing $n$ different balls. Show that the probability for every ball to be picked at least once in $d$ drawings is

$$1 - na_1^d + \binom{n}{2} a_2^d - \cdots + (-1)^n a_n^d$$

where

$$a_k \equiv a_k(n, m) = \binom{n-k}{m} \Big/ \binom{n}{m}$$

11. For two-index probability sequences $p_{jk}$, $j = 0, 1, \cdots$, $k = 0, 1, \cdots$ with probability generating function

$$P(t, u) = \sum_{j,k} p_{jk} t^j u^k$$

and binomial moments given by

$$B_{r,s} = \sum_{j,k} \binom{r}{j} \binom{k}{s} p_{jk}$$

show that the binomial moment generating function $B(t, u)$ is given by

$$B(t, u) = P(1 + t, 1 + u)$$

so that

$$p_{jk} = \sum_{r,s} (-1)^{r+s} \binom{j+r}{r} \binom{k+s}{s} B_{j+r,k+s}$$

TABLE 1

RENCONTRES NUMBERS $D_{n,k}$

| $n \backslash k$ | 0 | 1 | 2 | 3 | 4 | 5 | 6 | 7 | 8 | 9 | 10 |
|---|---|---|---|---|---|---|---|---|---|---|---|
| 0 | 1 | | | | | | | | | | |
| 1 | 0 | 1 | | | | | | | | | |
| 2 | 1 | 0 | 1 | | | | | | | | |
| 3 | 2 | 3 | 0 | 1 | | | | | | | |
| 4 | 9 | 8 | 6 | 0 | 1 | | | | | | |
| 5 | 44 | 45 | 20 | 10 | 0 | 1 | | | | | |
| 6 | 265 | 264 | 135 | 40 | 15 | 0 | 1 | | | | |
| 7 | 1854 | 1855 | 924 | 315 | 70 | 21 | 0 | 1 | | | |
| 8 | 14833 | 14832 | 7420 | 2464 | 630 | 112 | 28 | 0 | 1 | | |
| 9 | 133496 | 133497 | 66744 | 22260 | 5544 | 1134 | 168 | 36 | 0 | 1 | |
| 10 | 1334961 | 1334960 | 667485 | 222480 | 55650 | 11088 | 1890 | 240 | 45 | 0 | 1 |

CHAPTER 4

# The Cycles of Permutations

## 1. INTRODUCTION

Permutations may be regarded in two ways: (i) as ordered arrangements of given objects, as in Chapter 1, and (ii) as derangements of a standard order, usually taken as the natural (alphabetical or numerical) order, as in this chapter. These two ways are related as noun to verb, or as object to operator; the second is that naturally used in the study of permutations in the theory of groups.

To indicate completely a permutation of numbered elements as an operator, a notation like

$$\downarrow \begin{pmatrix} 12345 \\ 25431 \end{pmatrix}$$

is required, with the arrow indicating the direction of the operation, the first line the operand, and the second the result. This may be abbreviated by convention to the second line alone if both operand and direction are standardized, but another one-line notation, that by cycles, is of more interest here. In this notation the permutation above is written as

$$(125)(34)$$

this means that the operation replaces 1 by 2, 2 by 5, 5 by 1, and 3 by 4, 4 by 3. Each of these separate parts is called a cycle for obvious reasons.

It is clear that any permutation may be represented in this manner since a cycle starting with element 1 either includes all elements or does not, and, if it does not, the same argument applies to any element not in its cycle; so, eventually, every element is in some cycle. Moreover the representation is unique if it is agreed that expressions like (125), (251), and (512), which indicate the same cycle, are the same. It is conventional to write a cycle with its smallest element in the first position.

From this point of view a host of new enumeration problems appear: how many permutations have exactly $k$ unit cycles (that is, $k$ elements left unchanged), or $k$ $r$-cycles, or $k$ $r$-cycles or $s$-cycles, and so on? Or, if cycle length is ignored, how many permutations have $k$ cycles?

Again, the cycles may be supposed of some particular kind, such as those with elements in some specified sequence, and the same questions repeated, a subject extensively studied in a beautiful paper by Touchard, **8**.

A complete study could occupy a book, so attention is focussed here on salient results, and the reader is reminded that many particular results are contained in the problems.

## 2. CYCLE CLASSES

A permutation with $k_1$ unit cycles, $k_2$ 2-cycles, and so on, is said to be of cycle class $(k_1, k_2, \cdots)$ or, for brevity, of class $(k)$. Another notation is that of partitions:

$$1^{k_1}2^{k_2}\cdots$$

which has the same meaning. If $C(k_1, k_2, \cdots)$ is the number of permutations of $n$ elements of class $(k)$, so that

$$k_1 + 2k_2 + \cdots + nk_n = n$$

what is its formula?

To find this formula, pick an arbitrary permutation of class $(k)$ and permute its elements in all possible $(n!)$ ways. The resulting permutations are not distinct for just two reasons: (i) all cycles containing the same elements in the same cyclic order are the same, and (ii) the relative position of cycles is immaterial. In an $r$-cycle, as already noticed, there are $r$ possible initial elements and, hence, $r$ possible duplications for the first reason; the total number of such duplications is

$$1^{k_1}2^{k_2}\cdots n^{k_n}$$

Again, $k_r$ $r$-cycles may be permuted in $k_r!$ ways, so duplications for the second reason are enumerated by

$$k_1!\, k_2! \cdots k_n!$$

Hence,

$$C(k_1, k_2, \cdots, k_n) = \frac{n!}{1^{k_1}k_1!\, 2^{k_2}k_2! \cdots n^{k_n}k_n!} \tag{1}$$

with $k_1 + 2k_2 + \cdots nk_n = n$.

*Example 1.* The six permutations of 3 elements are divided into cycle classes as follows:

| *Class* | *Permutations* | *Number* |
|---|---|---|
| (001) | (123), (132) | 2 |
| (110) | (12)(3), (13)(2), (1)(23) | 3 |
| (300) | (1)(2)(3) | 1 |

In partition notation, the three classes are, respectively, 3, 21, $1^3$.

*Example 2.* The permutations of class $n$ (in partition notation), that is, with $k_n = 1$, $k_r = 0$, $r \neq n$, are given by

$$C(00 \cdots 01) = n!/n = (n-1)!$$

This is verified by the facts that in an $n$-cycle the first element is unity, by convention, and that there are $(n-1)!$ possible ways of assigning the remaining $n-1$ elements.

Also, the single permutation of class $1^n$ is readily identified as the identity permutation (in which no element is changed).

A generating function for the numbers $C(k_1, k_2, \cdots, k_n)$ must be a multiple generating function since $n$ kinds of cycle are to be tagged. It is defined by

$$C_n(t_1, t_2, \cdots, t_n) = \Sigma C(k_1, k_2, \cdots, k_n) t_1^{k_1} \cdots t_n^{k_n}$$

Hence, by equation (1)

$$C_n(t_1, t_2, \cdots, t_n) = \Sigma \frac{n!}{k_1! \cdots k_n!} \left(\frac{t_1}{1}\right)^{k_1} \left(\frac{t_2}{2}\right)^{k_2} \cdots \left(\frac{t_n}{n}\right)^{k_n} \qquad (2)$$

The sum is over all non-negative integral values of $k_1$ to $k_n$ such that $k_1 + 2k_2 + \cdots + nk_n = n$, or what is the same thing, over all partitions of $n$. $C_n(t_1, t_2, \cdots, t_n)$ is called the *cycle indicator* of the symmetric group.

The first few instances of (2), abbreviating $C_n(t_1, t_2, \cdots, t_n)$ to $C_n$, are

$$\begin{aligned} C_1 &= t_1 \\ C_2 &= t_1^2 + t_2 \\ C_3 &= t_1^3 + 3t_1t_2 + 2t_3 \\ C_4 &= t_1^4 + 6t_1^2t_2 + 3t_2^2 + 8t_1t_3 + 6t_4 \end{aligned}$$

and it is convenient to set $C_0 = 1$. Table 1 gives $C_n$ for $n = 1(1)9$.

By comparison of equation (2) and equation (2.46), it is apparent that

$$\begin{aligned} C_n(t_1, t_2, \cdots, t_n) &= A_n(1; t_1, t_2, 2!\, t_3, \cdots, (n-1)!\, t_n) \\ &= Y_n(t_1, t_2, 2!\, t_3, \cdots, (n-1)!\, t_n) \end{aligned} \qquad (3)$$

the last form by the unnumbered notational equation following equation (2.42*a*). By equation (2.45), this is the same as

$$\begin{aligned} \exp uC &= \sum_0^\infty C_n(t_1, t_2, \cdots, t_n) u^n/n! \\ &= \exp(ut_1 + u^2t_2/2 + u^3t_3/3 + \cdots) \end{aligned} \qquad (3a)$$

the basic generating identity, which is most convenient for later work. Further properties of the $C_n$ may be determined by specialization of results in Chapter 2. A few of these are worth noting now for verifications.

TABLE 1

CYCLE INDICATOR $C_n(t_1, t_2, \cdots, t_n)$

$$C_1 = t_1$$

$$C_2 = t_1^2 + t_2$$

$$C_3 = t_1^3 + 3t_1t_2 + 2t_3$$

$$C_4 = t_1^4 + 6t_1^2t_2 + 3t_2^2 + 8t_1t_3 + 6t_4$$

$$C_5 = t_1^5 + 10t_1^3t_2 + 15t_1t_2^2 + 20t_1^2t_3 + 20t_2t_3 + 30t_1t_4 + 24t_5$$

$$\begin{aligned} C_6 = t_1^6 &+ 15t_1^4t_2 + 45t_1^2t_2^2 + 40t_1^3t_3 + 15t_2^3 + 120t_1t_2t_3 + 90t_1^2t_4 \\ &+ 40t_3^2 + 90t_2t_4 + 144t_1t_5 + 120t_6 \end{aligned}$$

$$\begin{aligned} C_7 = t_1^7 &+ 21t_1^5t_2 + 105t_1^3t_2^2 + 70t_1^4t_3 + 105t_1t_2^3 + 420t_1^2t_2t_3 + 210t_1^3t_4 \\ &+ 210t_2^2t_3 + 280t_1t_3^2 + 630t_1t_2t_4 + 504t_1^2t_5 \\ &+ 420t_3t_4 + 504t_2t_5 + 840t_1t_6 + 720t_7 \end{aligned}$$

$$\begin{aligned} C_8 = t_1^8 &+ 28t_1^6t_2 + 210t_1^4t_2^2 + 112t_1^5t_3 + 420t_1^2t_2^3 + 1120t_1^3t_2t_3 + 420t_1^4t_4 \\ &+ 105t_2^4 + 1680t_1t_2^2t_3 + 1120t_1^2t_3^2 + 2520t_1^2t_2t_4 + 1344t_1^3t_5 \\ &+ 1120t_2t_3^2 + 1260t_2^2t_4 + 3360t_1t_3t_4 + 4032t_1t_2t_5 + 3360t_1^2t_6 \\ &+ 1260t_4^2 + 2688t_3t_5 + 3360t_2t_6 + 5760t_1t_7 + 5040t_8 \end{aligned}$$

$$\begin{aligned} C_9 = t_1^9 &+ 36t_1^7t_2 + 378t_1^5t_2^2 + 168t_1^6t_3 + 1260t_1^3t_2^3 + 2520t_1^4t_2t_3 + 756t_1^5t_4 \\ &+ 945t_1t_2^4 + 7560t_1^2t_2^2t_3 + 3360t_1^3t_3^2 + 7560t_1^3t_2t_4 + 3024t_1^4t_5 \\ &+ 2520t_2^3t_3 + 10080t_1t_2t_3^2 + 11340t_1t_2^2t_4 + 15120t_1^2t_3t_4 + 18144t_1^2t_2t_5 \\ &+ 10080t_1^3t_6 + 2240t_3^3 + 15120t_2t_3t_4 + 9072t_2^2t_5 + 11340t_1t_4^2 \\ &+ 24192t_1t_3t_5 + 30240t_1t_2t_6 + 25920t_1^2t_7 \\ &+ 18144t_4t_5 + 20160t_3t_6 + 25920t_2t_7 + 45360t_1t_8 + 40320t_9 \end{aligned}$$

First, since every permutation belongs in a cycle class, the sum of the numerical coefficients of $C_n$ is $n!$, that is,

$$C_n(1, 1, \cdots, 1) = n! \tag{4}$$

This is consistent with (3*a*) since

$$\begin{aligned} \exp(u + u^2/2 + u^3/3 + \cdots) &= \exp\log(1 - u)^{-1} \\ &= (1 - u)^{-1} \\ &= 1 + u + u^2 + \cdots \end{aligned}$$

and it may be noticed that, by (2), equation (4) also may be written

$$\Sigma \frac{1}{1^{k_1}k_1!\, 2^{k_2}k_2! \cdots n^{k_n}k_n!} = 1$$

which is a famous identity given by Cauchy.

Next, from equations (2.44) and (3), or by differentiation of (3$a$) with respect to $u$,

$$C_{n+1} = \sum_{0}^{n} (n)_k t_{k+1} C_{n-k} \tag{5}$$

The arguments of all $C$'s have been omitted. Equation (5) may be checked by the instances of equation (2) already given or by Table 1; an additional example is as follows:

$$\begin{aligned} C_5 &= t_1 C_4 + 4t_2 C_3 + 12t_3 C_2 + 24t_4 C_1 + 24t_5 C_0 \\ &= t_1^5 + 10t_1^3 t_2 + 15t_1 t_2^2 + 20t_1^2 t_3 + 20t_2 t_3 + 30t_1 t_4 + 24t_5 \end{aligned}$$

Finally, for further checks on Table 1, note that

$$r\frac{d}{dt_r}(\exp uC) = u^r \exp uC$$

and hence

$$r\frac{dC_n}{dt_r} = (n)_r C_{n-r} \tag{6}$$

Thus, for example,

$$\begin{aligned} \frac{dC_5}{dt_1} &= 5t_1^4 + 30t_1^2 t_2 + 15t_2^2 + 40t_1 t_3 + 30t_4 = 5C_4 \\ \frac{dC_n}{dt_n} &= (n-1)! \\ \frac{dC_n}{dt_{n-1}} &= n(n-2)!t_1 \end{aligned}$$

## 3. PERMUTATIONS BY NUMBER OF CYCLES

How many permutations are there with $k$ cycles, regardless of their lengths? By the development above, the answer is given by the generating function

$$C_n(t, t, \cdots, t) \equiv c_n(t)$$

that is, by setting each $t_r$ equal to $t$.

Hence by (3$a$)

$$\begin{aligned} \exp uc(t) &= \exp t(u + u^2/2 + u^3/3 + \cdots) \\ &= \exp t \log (1-u)^{-1} \\ &= (1-u)^{-t} \\ &= 1 + \sum_{1}^{\infty} t(t+1)\cdots(t+n-1)\frac{u^n}{n!} \end{aligned} \tag{7}$$

and $c_0(t) = 1$,

$$c_n(t) = t(t + 1) \cdots (t + n - 1), \qquad n > 0 \tag{8}$$

This is the generating function of the signless Stirling numbers of the first kind, $(-1)^{k+n}s(n, k)$ of Section 2.7; thus, if

$$c_n(t) = \sum_0^n c(n, k)t^k$$

then

$$c(n, k) = (-1)^{k+n}s(n, k) \tag{9}$$

These are the numbers asked for. Direct proof, which serves as a verification of the results above, is as follows.

Consider the permutations of $n$ elements with $k$ cycles, enumerated by $c(n, k)$, according to the condition of the last element $n$. Element $n$ is either in a unit cycle or it is not. If it is, it is associated with $n - 1$ elements in $k - 1$ cycles, enumerated by $c(n - 1, k - 1)$. If it is not, the number of permutations is the number of ways element $n$ may be added to the permutations of $n - 1$ elements with $k$-cycles without creating a new cycle. This number is $n - 1$ since, in each cycle of length $r$, there are exactly $r$ possible places for element $n$ (note that the leading position is excluded because $n$ is the largest element) and by assumption the sum of the cycle lengths is $n - 1$. Hence

$$c(n, k) = c(n - 1, k - 1) + (n - 1)c(n - 1, k) \tag{10}$$

This corresponds to the generating-function relation

$$c_n(t) = (t + n - 1)c_{n-1}(t) \tag{11}$$

and, since $c_1(t) = t$, to

$$c_n(t) = t(t + 1) \cdots (t + n - 1)$$

as before.

The recurrence relation (10) naturally agrees with equation (2.36).

Now consider the same problem as a probability problem. The ratio $c(n, k)/n!$ is the probability that a permutation picked at random has $k$ cycles. Hence the generating function of probabilities is $c_n(t)/n!$, and the factorial moment generating function, or the binomial moment generating function, by Section 2.6, is $c_n(1 + t)/n!$. But, by equation (8)

$$c_n(1 + t) = t^{-1}c_{n+1}(t)$$

and the $k$th moments are

$$(m)_k = k!B_k = k!\, c(n + 1, k + 1)/n! \tag{12}$$

The first two instances of (12) are

$$m_0 = B_0 = c(n + 1, 1)/n! = 1$$

$$m_1 = B_1 = c(n + 1, 2)/n! = 1 + \frac{1}{2} + \cdots + \frac{1}{n}$$

Note that the mean number of cycles, $m_1$, approaches $\log n$ for large values of $n$; hence, the mean length of cycles approaches $n/\log n$.

For higher moments, it is easier to work with a recurrence relation on $n$, and it is convenient, purely notationally, to work with binomial moments. Write $B_k(n)$ for the $k$th binomial moment, and

$$B(t, n) = \Sigma B_k(n)t^k = c_n(1 + t)/n!$$

for its generating function. Then, by (11)

$$B(t, n) = (1 + t/n)B(t, n - 1) \tag{13}$$

and

$$B_k(n) = B_k(n - 1) + n^{-1}B_{k-1}(n - 1) \tag{14}$$

This recurrence relation may be solved for successive values of $k$. Thus, since $B_0(n) = 1$ and $B_1(1) = 1$,

$$\begin{aligned} B_1(n) &= B_1(n - 1) + n^{-1} \\ &= B_1(n - 2) + (n - 1)^{-1} + n^{-1} \\ &= 1 + \frac{1}{2} + \cdots + \frac{1}{n} \end{aligned}$$

Again, from the equation for the variance $V(n)$, namely,

$$V(n) = 2B_2(n) + B_1(n) - B_1^2(n)$$

and the instances $k = 1, 2$ of (14), it follows that

$$\begin{aligned} V(n) &= V(n - 1) + (n - 1)/n^2 \\ &= V(n - 2) + (n - 2)/(n - 1)^2 + (n - 1)/n^2 \qquad (15) \\ &= \frac{1}{4} + \frac{2}{9} + \cdots + \left(\frac{n - 1}{n^2}\right) \end{aligned}$$

which again approaches $\log n$ for $n$ large. Both these results agree with those obtained quite differently by Feller, **1**, p. 205, et seq., which the reader should compare.

## 4. PERMUTATIONS WITHOUT UNIT CYCLES

How many permutations of $n$ elements have $k$ cycles, no one of which is a unitary cycle?

The answer, by definition of $C_n$, is clearly the coefficient of $t^k$ in $C_n(0, t, t, \cdots, t)$ or for brevity $d_n(t)$. By equation (3$a$)

$$\exp ud(t) = \exp t(u^2/2 + u^3/3 + \cdots) \qquad (16)$$
$$= (1 - u)^{-t} \exp(-tu)$$

and, hence, by comparison with (7)

$$d_n(t) = \sum_0^n \binom{n}{k} c_{n-k}(t)(-t)^k \qquad (17)$$

with $c_0(t) = 1$, $c_n(t) = t(t + 1) \cdots (t + n - 1)$, $n > 0$, as in (8).

Then, if

$$d_n(t) = \Sigma d(n, k)t^k$$

the coefficient relation corresponding to (17) is

$$d(n, k) = \sum_0^n \binom{n}{j} (-1)^j c(n - j, k - j) \qquad (18)$$

because of which the numbers $d(n, k)$ are called *associated Stirling numbers of the first kind.* It is worth noticing that the companion relation to (17), obtained by multiplying (16) through by $e^{tu}$ and equating coefficients of $u^n$, is

$$c_n(t) = \Sigma \binom{n}{j} d_{n-j}(t)t^j$$

which corresponds to

$$c(n, k) = \Sigma \binom{n}{j} d(n - j, k-j) \qquad (19)$$

an interesting representation of the Stirling numbers. Finally

$$d_n(1) = \Sigma d(n, k) = D_n$$

with $D_n$ a rencontres number, follows at once from (16) and equation (3.23). Table 2 shows the numbers $d(n, k)$ for $n = 0(1)10$.

The answer is now formally complete, but simpler relations exist. Thus, by differentiation of (16) with respect to $u$, it follows, with the usual convention $(d(t))^n \equiv d_n(t)$, that

$$d(t) \exp ud(t) = -t \exp ud(t) + t(1 - u)^{-1} \exp ud(t) \qquad (20)$$

or

$$(1 - u)d(t) \exp ud(t) = tu \exp ud(t)$$

This corresponds to

$$d_{n+1}(t) = nd_n(t) + ntd_{n-1}(t) \qquad (21)$$

which in its turn corresponds to

$$d(n + 1, k) = nd(n, k) + nd(n - 1, k - 1) \tag{22}$$

a simple recurrence relation which may be verified by direct argument as well as by the numbers in Table 2.

Indeed, in the permutations of $n + 1$ elements which have $k$ cycles, no one of which is a unit cycle, enumerated by $d(n + 1, k)$, element $n + 1$ is in some 2-cycle or it is not. If it is, by the cycle-ordering convention, it is last and is paired with one of the other $n$ elements, the remaining $n - 1$ elements may appear in $k - 1$ cycles in $d(n - 1, k - 1)$ ways, and the total number of such permutations is $nd(n - 1, k - 1)$. If it is not, it is inserted into one of the $k$ cycles of $n$ elements enumerated by $d(n, k)$, and this may be done in $n$ ways. So each term of (22) is verified.

It is clear that the presence or absence of cycles of lengths other than unity is included in the cycle indicator with equal ease, but, of course, the mathematical development to obtain numerical results becomes more extensive as more elaborate enumeration is asked for. Further illustration appears in the problems.

## 5. ENUMERATION BY CYCLE CHARACTER

In the material above, cycles are distinguished only by their lengths (the number of elements they contain). It is possible to go further by imparting a character to the ordering of elements within a cycle. This is done here with the single restriction that the ordering character for any cycle be independent of the particular choice of elements for that cycle.

Any particular cycle of $r$ has $(r - 1)!$ orderings, since by the convention the smallest must be first. Each of these is a candidate for a cycle character and may be indicated by a variable $t_{ri}$, $i = 1$ to $(r - 1)!$. The cycle indicator accounting for all orderings is a function of $N = 1 + 1 + 2 + \cdots (n - 1)!$ variables, and is a refinement of the indicator appearing above. The simplest examples are

$$C_3(t_1, t_2, t_{31}, t_{32}) = t_1^3 + 3t_1t_2 + t_{31} + t_{32}$$

$$\begin{aligned} C_4(t_1, t_2, t_{31}, t_{32}, t_{41}, \cdots, t_{46}) &= t_1^4 + 6t_1^2t_2 + 3t_2^2 + 4t_1(t_{31} + t_{32}) \\ &\quad + t_{41} + t_{42} + t_{43} + t_{44} + t_{45} + t_{46} \end{aligned}$$

Note that the function is symmetrical in $t_{ri}$ for each $r$, and that the former indicator is restored by setting each $t_{ri}$ equal to $t_r$.

The cycle indicator for a single selected ordering for every length of cycle, of course not necessarily in a similar way for each length, is obtained by setting for each $r$ all $t_{ri}$ save one to zero. If the single one remaining is relabeled $s_r$, the ordered-cycle indicator may be written

$$A_n(s_1, s_2, \cdots, s_n) = \Sigma \frac{n!}{k_1! \cdots k_n!} \left(\frac{s_1}{1!}\right)^{k_1} \cdots \left(\frac{s_n}{n!}\right)^{k_n} \tag{23}$$

$$= C_n(s_1, s_2, s_3/2, \cdots, s_n/(n-1)!)$$

As indicated by the last line, this is obtained from equation (2) by replacing $t_r$ by $s_r/(r-1)!$, an operation justified by the remarks at the end of the preceding paragraph.

Note that by (3)

$$A_n(s_1, s_2, \cdots, s_n) = Y_n(s_1, s_2, \cdots, s_n) \tag{23a}$$

with $Y_n$ being the Bell polynomial of Chapter 2; hence Table 2.3, apart from notation and with all $f_i$ set at unity, is a table for this ordered-cycle indicator.

TABLE 2

ASSOCIATED STIRLING NUMBERS OF THE FIRST KIND

$d(n, k)$

| $n\backslash k$ | 0 | 1 | 2 | 3 | 4 | 5 |
|---|---|---|---|---|---|---|
| 0 | 1 | | | | | |
| 1 | 0 | 0 | | | | |
| 2 | 0 | 1 | | | | |
| 3 | 0 | 2 | | | | |
| 4 | 0 | 6 | 3 | | | |
| 5 | 0 | 24 | 20 | | | |
| 6 | 0 | 120 | 130 | 15 | | |
| 7 | 0 | 720 | 924 | 210 | | |
| 8 | 0 | 5040 | 7308 | 2380 | 105 | |
| 9 | 0 | 40320 | 64224 | 26432 | 2520 | |
| 10 | 0 | 362880 | 623376 | 303660 | 44100 | 945 |

Enumerations of permutations with ordered cycles of specified character may now be made from the ordered-cycle indicator exactly as above. Note that the correspondent to (3a), with $A^n \equiv A_n(s_1, s_2, \cdots, s_n)$ is

$$\exp uA = \exp (us_1 + u^2 s_2/2! + u^3 s_3/3! + \cdots) \tag{23b}$$

For illustration, consider the enumeration by number of ordered cycles. The enumerator is $A_n(s, \cdots, s) \equiv a_n(s)$, and by (23$b$)

$$\begin{aligned} \exp ua(s) &= \exp s(u + u^2/2! + u^3/3! + \cdots) \\ &= \exp s(e^u - 1) \qquad (24) \\ &= \sum_0^\infty s^k(e^u - 1)^k/k! = \sum_{n=0}^\infty (u^n/n!) \sum_{k=0}^n S(n, k)s^k \end{aligned}$$

the last step by Problem 2.14 [with $S(n, k)$ the Stirling number of the second kind]. Hence

$$a_n(s) = \Sigma S(n, k)s^k \qquad (25)$$

The first few values are

$$\begin{aligned} a_0(s) &= 1 \qquad & a_2(s) &= s + s^2 \\ a_1(s) &= s & a_3(s) &= s + 3s^2 + s^3 \end{aligned}$$

which may be verified from Table 2.3.

The Stirling numbers $S(n, k)$ then enumerate the permutations of $n$ elements with $k$ cycles when only those cycles with a specified order are

TABLE 3

ASSOCIATED STIRLING NUMBERS OF THE SECOND KIND

$b(n, k)$

| $n \backslash k$ | 0 | 1 | 2 | 3 | 4 | 5 |
|---|---|---|---|---|---|---|
| 0 | 1 | | | | | |
| 1 | 0 | 0 | | | | |
| 2 | 0 | 1 | | | | |
| 3 | 0 | 1 | | | | |
| 4 | 0 | 1 | 3 | | | |
| 5 | 0 | 1 | 10 | | | |
| 6 | 0 | 1 | 25 | 15 | | |
| 7 | 0 | 1 | 56 | 105 | | |
| 8 | 0 | 1 | 119 | 490 | 105 | |
| 9 | 0 | 1 | 246 | 1918 | 1260 | |
| 10 | 0 | 1 | 501 | 6825 | 9450 | 945 |

permitted; this is abstractly identical with a distribution problem appearing in the next chapter (Section 5.6), namely, the number of ways $n$ different things may be put into $k$ like cells with no cells empty, for which the same answer is obtained.

The recurrence relation for $a_n(s)$, which follows by differentiation of (24) with respect to $u$, is

$$a_{n+1}(s) = s[a(s) + 1]^n, \qquad a^n(s) \equiv a_n(s) \qquad (26)$$

Actual computation of the polynomials by this recurrence may be made rapidly, using only additions, by forming the triangular array below (the idea is due to Aitken*).

$$\begin{array}{llll}
a_0(s) & & & \\
a_1(s) & a_0(s)+a_1(s) & & \\
a_2(s) & a_1(s)+a_2(s) & a_0(s)+2a_1(s)+a_2(s) & \\
a_3(s) & a_2(s)+a_3(s) & a_1(s)+2a_2(s)+a_3(s) & \\
 & & & a_0(s)+3a_1(s)+3a_2(s)+a_3(s)
\end{array}$$

The array is formed as follows: the entry in the first column, by (26), is $s$ times the entry farthest right in the preceding row, the entry in any other row and column is the sum of two entries in the column to the left, one in the same row, the other in the preceding row. The array is even simpler for the numbers $a_n \equiv a_n(1)$ which have the recurrence

$$a_{n+1} = (a+1)^n, \qquad a^k \equiv a_k \tag{26a}$$

For the enumeration by number of cycles, no one of which is a unit cycle, the number is $A_n(0, s, \cdots, s) = b_n(s)$, and

$$\begin{aligned} \exp ub(s) &= \exp s(e^u - 1 - u) \\ &= \exp u(a(s) - s) \end{aligned} \tag{27}$$

Hence

$$b_n(s) = \Sigma \binom{n}{k} a_{n-k}(s)(-s)^k \tag{28}$$

$$a_n(s) = \Sigma \binom{n}{k} b_{n-k}(s)s^k \tag{29}$$

which imply relationships of the coefficients $b(n, k)$ in

$$b_n(s) = \Sigma b(n, k)s^k$$

to $S(n, k)$ similar to those of $d(n, k)$ to $c(n, k)$ in Section 4. The numbers $b(n, k)$ may be called *associated Stirling numbers of the second kind.*

The recurrence relation for $b_n(s)$ corresponding to (26) is

$$b_{n+1}(s) = s[(b(s)+1)^n - b_n(s)], \qquad b^k(s) \equiv b_k(s) \tag{30}$$

which is also easily computed by the Aitken triangular array. For the

* A. C. Aitken, A problem on combinations, *Math. Notes Edinburgh*, vol. 21 (1933), pp. 18–23.

numbers $b(n, k)$ a second recurrence is more convenient. This follows from the generating function relation

$$\left(\frac{\partial}{\partial u} - s\frac{\partial}{\partial s}\right) \exp ub(s) = su \exp ub(s) \tag{31}$$

Hence with a prime denoting a derivative

$$b_{n+1}(s) = sb_n'(s) + snb_{n-1}(s) \tag{32}$$

and

$$b(n + 1, k) = kb(n, k) + nb(n - 1, k - 1) \tag{33}$$

Table 3 gives $b(n, k)$ for $n = 0(1)10$.

Finally it should be noticed that the complete cycle indicator with double suffix variables $t_{ri}$ may be used to obtain cycle indicators other than $A_n$ and $C_n$. Simply to illustrate further possibilities, it may be mentioned that, if a fixed order is specified for merely the first two elements of a cycle of $r(r > 2)$, the cycle indicator becomes

$$\begin{aligned} A_n^*(q_1, q_2, \cdots, q_n) &= \Sigma \frac{n!}{k_1! \cdots k_n!}\left(\frac{q_1}{1}\right)^{k_1}\left(\frac{q_2}{2}\right)^{k_2}\left(\frac{q_3}{3 \cdot 2}\right)^{k_3}\cdots\left(\frac{q_n}{n(n-1)}\right)^{k_n} \\ &= C_n(q_1, q_2, \cdots, q_n/(n-1)) \end{aligned} \tag{34}$$

## 7. CYCLES OF EVEN AND ODD PERMUTATIONS

A permutation is even if it is equivalent to an even number of transpositions, and odd otherwise. Every cycle of odd order corresponds to an even number of transpositions, and cycles of even order to an odd number of transpositions. Hence the class of even permutations (the alternating group) consists of those having an even number of cycles of even order, the class of odd permutations of those having an odd number of cycles of even order.

The cycle indicator for even permutations is then the mean of the indicator for all permutations and the same indicator with negative signs for all indexes of even cycles. This is independent of other properties supposed for the cycles (the same rule goes for $C_n$, $A_n$ and $A_n^*$), and it is helpful to have a notation which emphasizes this. Thus if $C_n^e(t_1, t_2, \cdots, t_n)$ is the indicator of even permutations for cycles apart from order, and $C_n^o(t_1, \cdots, t_n)$ is the same quantity for odd permutations, then

$$2C_n^e(t_1, t_2, \cdots, t_n) = C_n(t_1, t_2, \cdots, t_n) + C_n(t_1, -t_2, t_3, -t_4, \cdots) \tag{35}$$

$$2C_n^o(t_1, t_2, \cdots, t_n) = C_n(t_1, t_2, \cdots, t_n) - C_n(t_1, -t_2, t_3, -t_4, \cdots) \tag{36}$$

The first few values are $C_0^e = 1$, $C_0^o = 0$, and

$$C_1^e = t_1 \qquad C_2^e = t_1^2 \qquad C_3^e = t_1^3 + 2t_3 \qquad C_4^e = t_1^4 + 3t_2^2 + 8t_1t_3$$

$$C_1^o = 0 \qquad C_2^o = t_2 \qquad C_3^o = 3t_1t_2 \qquad C_4^o = 6t_1^2t_2 + 6t_4$$

From these equations, or their correspondents for ordered cycles, any of the enumerations made for all permutations may be repeated for even and odd permutations. For illustration, if the enumerators for even and odd permutations by number of cycles are $c_n^e(t)$ and $c_n^o(t)$, then

$$c_n^e(t) = C_n^e(t, t, \cdots, t)$$

$$c_n^o(t) = C_n^o(t, t, \cdots, t)$$

and by (35) and (36)

$$2 \exp uc^e(t) = (1 - u)^{-t} + (1 + u)^t \tag{37}$$

$$2 \exp uc^o(t) = (1 - u)^{-t} - (1 + u)^t \tag{38}$$

From these, it is found that

$$2c_n^e(t) = t(t + 1) \cdots (t + n - 1) + t(t - 1) \cdots (t - n + 1) \tag{39}$$

$$2c_n^o(t) = t(t + 1) \cdots (t + n - 1) - t(t - 1) \cdots (t - n + 1) \tag{40}$$

Further development of these functions appears in the problems.

## REFERENCES


1. W. Feller, *An Introduction to Probability Theory and its Applications*, New York, 1950.
2. S. Chowla, I. N. Herstein, and K. Moore, On recursions connected with symmetric groups I, *Canadian Journal of Math.*, vol. 3 (1951), pp. 328–334.
3. S. Chowla, I. N. Herstein, and W. R. Scott, The solutions of $x^d = 1$ in symmetric groups, *Norske Vid. Selsk. Forh. Trondheim*, vol. 25 (1952), pp. 29–31.
4. W. Gontcharoff, Sur la distribution des cycles dans les permutations, *C.R. (Doklady) Acad. Sci. URSS (N.S.)*, vol. 35 (1942), pp. 267–269.
5. E. Jacobsthal, Sur le nombre d'éléments du group symétrique $S_n$ dont l'ordre est un nombre premier, *Norske Vid. Selsk. Forh. Trondheim*, vol. 21 (1949), pp. 49–51.
6. L. Moser and M. Wyman, On solutions of $x^d = 1$ in symmetric groups, *Canadian Journal of Math.*, vol. 7 (1955), pp. 159–168.
7. J. Touchard, Propriétés arithmétiques de certains nombres recurrents, *Ann. Soc. Sci. Bruxelles*, vol. A53 (1933), pp. 21–31.
8. ———, Sur les cycles des substitutions, *Acta Math.*, vol. 70 (1939), pp. 243–279.
9. ———, Nombres exponentials et nombres de Bernoulli, *Canadian Journal of Math.*, vol. 8 (1956), pp. 305–320.

## PROBLEMS

1. In the abbreviated notation of Problem 2.21, write the cycle indicator $C_n(t_1, \cdots, t_n)$ as $C_n(t)$, $C_n(s_1 + t_1, \cdots, s_n + t_n)$ as $C_n(s + t)$; then by Problem 2.21,

$$C_n(s + t) = (C(s) + C(t))^n, \qquad C^k(s) = C_k(s_1, \cdots, s_k)$$

Using this result and

$$C_n(x, x^2, \cdots, x^n) = n!\, x^n$$

show that

$$\begin{aligned} C_n(1 + x, 1 + x^2, \cdots, 1 + x^n) &= n!\,(1 + x + \cdots + x^n) \\ &= n!\,(1 - x^{n+1})/(1 - x) \end{aligned}$$

$$C_n(1 + x + x^2, 1 + x^2 + x^4, \cdots, 1 + x^n + x^{2n}) = n!\,\frac{1 - x^{n+1}(1 + x) + x^{2n+3}}{(1 - x)(1 - x^2)}$$

and verify by use of (3*a*).

2. (*a*) For the ordered-cycle indicator $A_n(s_1, \cdots, s_n)$, and with

$$A_n(1, 1, \cdots, 1) = a_n(1) = a_n$$

$$A_n(x, x^2, \cdots, x^n) = a_n x^n$$

show similarly that

$$A_n(1 + x, \cdots, 1 + x^n) = (a + ax)^n, \qquad a^k \equiv a_k$$

Abbreviating $A_n(1 + x, \cdots, 1 + x^n)$ to $A_n(x)$, verify (by Table 2.3) the initial values

$$A_0(x) = 1 \qquad A_2(x) = 2 + 2x + 2x^2$$

$$A_1(x) = 1 + x \qquad A_3(x) = 5 + 6x + 6x^2 + 5x^3$$

$$A_4(x) = 15 + 20x + 24x^2 + 20x^3 + 15x^4$$

$$A_5(x) = 52 + 75x + 100x^2 + 100x^3 + 75x^4 + 52x^5$$

(*b*) Derive the relations (the prime denotes a derivative)

$$A_n'(x) = n(A + x)^{n-1}, \qquad A^k(x) = A_k(x) \equiv A_k(1 + x, \cdots, 1 + x^k)$$

$$A_n(1) = 2(A + 1)^{n-1}, \qquad A^k \equiv A_k(1)$$

$$A_n'(1) = nA_n(1)/2$$

3. Show that the enumerators of permutations by number of cycles, $c_n(t)$ of Section 3, satisfy the congruence

$$c_{n+m}(t) \equiv c_n(t)c_m(t), \qquad (\text{mod } n)$$

*Hint*: Use the recurrence $c_{n+1}(t) = (n + t)c_n(t)$ and mathematical induction. For $n = p$, a prime, more explicit results may be obtained from Lagrange's identical congruence

$$(t)_p = t(t - 1) \cdots (t - p + 1) \equiv t^p - t, \qquad (\text{mod } p)$$

which is a statement that all Stirling numbers of the first kind, $s(p, k)$, are divisible by $p$ except $s(p, p)$ which is unity and $s(p, 1)$ which is $(p-1)!$ and has a remainder of $-1$ when divided by $p$, a result known as Wilson's theorem. Since

$$c_p(t) = (-1)^p(-t)_p$$

it follows that

$$c_p(t) \equiv t^p - t, \qquad (\text{mod } p)$$

4. For the enumerator $d_n(t)$ of Section 4, show similarly that

$$d_{n+m}(t) \equiv d_n(t)d_m(t), \qquad (\text{mod } n)$$

and, with $p$ again a prime,

$$d_p(t) \equiv -t, \qquad (\text{mod } p)$$

Verify these results for $p = 3$ and 5 by Table 2.

*Hint*: for the latter result, use (17), the congruence for $c_p(t)$ and

$$(1 + t)^p \equiv 1 + t^p, \qquad (\text{mod } p)$$

5. (*a*) For the enumerators $a_n(s)$ of permutations by number of ordered cycles, Section 5, show that

$$(a(s))_n = a(s)(a(s) - 1) \cdots (a(s) - n + 1) = s^n, \qquad a^k(s) \equiv a_k(s)$$

*Hint*: make the substitution $t = e^u - 1$ in equation (24).

(*b*) Using Lagrange's congruence, as given in Problem 3, and

$$(t)_{p+k} \equiv (t)_p(t)_k, \qquad (\text{mod } p)$$

show that

$$(t^p - t)t^k \equiv \sum_{j=0}^{k} S(k, j)(t)_{p+j}, \qquad (\text{mod } p)$$

where $S(k, j)$ is a Stirling number.

(*c*) Using the results of 5(*a*) and 5(*b*), derive

$$a_p(s) \equiv s^p + s, \qquad (\text{mod } p)$$

$$a_{p+k}(s) \equiv a_{k+1}(s) + s^p a_k(s), \qquad (\text{mod } p) \qquad \text{(Touchard, 7)}$$

and from these, with $a_n \equiv a_n(1)$,

$$a_p \equiv 2, \qquad (\text{mod } p)$$

$$a_{p+k} \equiv a_{k+1} + a_k, \qquad (\text{mod } p)$$

6. For the enumerators $b_n(s)$ of Section 5, show similarly that

$$(b(s))_n = b(s)(b(s) - 1) \cdots (b(s) - n + 1), \qquad b^k(s) \equiv b_k(s)$$

$$= (s - \beta(s))^n, \qquad \beta^k(s) \equiv s(s + 1) \cdots (s + k - 1)$$

$$b_p(s) \equiv a_p(s) - s^p \equiv s, \qquad (\text{mod } p)$$

$$b_{p+k}(s) \equiv b_{k+1}(s) + sb_k(s), \qquad (\text{mod } p)$$

7. From equation (29) or otherwise, derive the relation

$$S(n, k) = \sum \binom{n}{j} b(n - j, k - j)$$

or

$$S(n, n - k) = \sum_{j=0}^{n-k} \binom{n}{j + k} b(j + k, j)$$

Verify the instances

$$S(n, n - 1) = \binom{n}{2}$$

$$S(n, n - 2) = \binom{n}{3} + 3\binom{n}{4}$$

$$S(n, n - 3) = \binom{n}{4} + 10\binom{n}{5} + 15\binom{n}{6}$$

8. Write $C(n, r, k)$ for the number of permutations of $n$ elements having $k$ cycles of length $r$, the character of cycles not of length $r$ being ignored, and $C_n(r, t) = \Sigma C(n, r, k)t^k$ for its enumerator. Then

$$C_n(r, t) = C_n(1, 1, \cdots, t, 1, \cdots, 1)$$

with $C_n(t_1, \cdots, t_n)$ the cycle indicator and $t$ in the $r$th position. Derive the results

$$(1 - u) \exp uC(r, t) = \exp (t - 1)u^r/r, \qquad C^k(r, t) \equiv C_k(r, t)$$

$$C_n(r, t) - nC_{n-1}(r, t) = (t - 1)^k n!/r^k k!, \qquad n = kr$$

$$= 0, \qquad n \neq kr$$

Verify the instances tabled below

| $n$ | 1 | 2 | 3 | 4 |
|---|---|---|---|---|
| $C_n(1, t)$ | $t$ | $1 + t^2$ | $2 + 3t + t^3$ | $9 + 8t + 6t^2 + t^4$ |
| $C_n(2, t)$ | 1 | $1 + t$ | $3 + 3t$ | $15 + 6t + 3t^2$ |
| $C_n(3, t)$ | 1 | 2 | $4 + 2t$ | $16 + 8t$ |

Note that $C_n(1, t) \equiv D_n(t)$, the displacement number polynomial of Section 3.5; compare Problem 3.6.

9. (*a*) To simplify Problem 8, write

$$C_{kr}(r, t) = (kr)!\, c_k(r, t)/r^k k!$$

Derive the recurrence

$$c_k(r, t) = krc_{k-1}(r, t) + (t - 1)^k$$

$$= (kr - 1 + t)c_{k-1}(r, t) + (1 - t)(k - 1)rc_{k-2}(r, t)$$

For $r = 1$, $C_k(1, t) = c_k(1, t) = D_k(t)$; for $r = 2$, verify the values (by the table of Problem 8); $c_1(2, t) = 1 + t$, $c_2(2, t) = C_4(2, t)/3 = 5 + 2t + t^2$.

(*b*) Iterate the first form of the recurrence for $c_k(r, t)$ to show that

$$c_k(r, t) = \sum_0^k \binom{k}{j} r^j j!\, (t - 1)^{k-j}$$

$$= (\gamma(r) + t - 1)^k, \qquad \gamma^j(r) \equiv \gamma_j(r) = r^j j!$$

Hence

$$c_k(r, 0) = [\gamma(r) - 1]^k$$

and

$$c_k(r, t) = [c(r, 0) + t]^k, \qquad c^j(r, 0) \equiv c_j(r, 0)$$

Verify the table

| $n$ | 0 | 1 | 2 | 3 | 4 | 5 |
|---|---|---|---|---|---|---|
| $c_n(1, 0)$ | 1 | 0 | 1 | 2 | 9 | 44 |
| $c_n(2, 0)$ | 1 | 1 | 5 | 29 | 233 | 2329 |
| $c_n(3, 0)$ | 1 | 2 | 13 | 116 | 1393 | 20894 |

(*c*) Derive the formula

$$C_n(r, t) = \sum_0^m \frac{n!}{j!\, r^j}(t - 1)^j, \qquad m = [n/r]$$

10. To verify the results above for $r = 2$ by the method of inclusion and exclusion, show that, in the notation of Chapter 3,

$$n!\, S_k = \frac{1}{k!}\binom{n}{2}\binom{n-2}{2}\cdots\binom{n-2k+2}{2}(n - 2k)!$$
$$= n!/2^k k!$$

and, hence, that

$$C_n(2, t) = \sum_0^m n!\,(t - 1)^k/2^k k!, \qquad m = [n/2]$$

so that

$$C_{2n}(2, t) = 2nC_{2n-1}(2, t) + (t - 1)^n(2n)!/2^n n$$
$$C_{2n+1}(2, t) = (2n + 1)C_{2n}(2, t)$$

11. (*a*) From the first recurrence of Problem 9(*a*), determine the exponential generating function relation

$$\exp uc(r, t) = ru \exp uc(r, t) + \exp u(t - 1)$$

Hence

$$\exp u[c(r, t) + 1 - t] = (1 - ru)^{-1}$$

and

$$[c(r, t) + 1 - t]^n = r^n n! = \gamma_n(r)$$

with $c^k(r, t) \equiv c_k(r, t)$ and $\gamma_n(r)$ as in Problem 9(*b*).

(*b*) From the last relation above, show that, with $p$ a prime,

$$c_p(r, t) + 1 - t^p \equiv 0, \qquad (\text{mod } p)$$

$$c_{p+k}(r, t) + (1 - t^p)c_k(r, t) \equiv 0, \qquad (\text{mod } p)$$

and in particular

$$c_p(r, 0) + 1 \equiv 0, \qquad (\text{mod } p)$$
$$c_{2p+k}(r, 0) \equiv c_k(r, 0), \qquad (\text{mod } p)$$

Verify the following table of residues ($r = 2$)

| $n$ | 0 | 1 | 2 | 3 | 4 | 5 | 6 | 7 | |
|---|---|---|---|---|---|---|---|---|---|
| $c_n(2, 0)$ | 1 | 1 | −1 | −1 | −1 | 1 | 1 | 1 | (mod 3) |
| $c_n(2, 0)$ | 1 | 1 | 0 | −1 | −2 | −1 | −1 | 0 | (mod 5) |

12. Using the equation of Problem 9(*c*), show that the factorial moments of the distribution $C_n(r, t)/n!$ are given by

$$\begin{aligned}(m)_k &= r^{-k}, \qquad k \leq [n/r] \\ &= 0, \qquad k > [n/r]\end{aligned}$$

Hence the distribution is almost Poisson with mean $r^{-1}$ (see Gontcharoff, **4**).

13. Write $C(n, r, s, j, k)$ for the number of permutations having $j$ $r$-cycles and $k$ $s$-cycles, and

$$C_n(r, s; t, w) = \Sigma\Sigma C(n, r, s, j, k)t^j w^k$$

for their enumerator. Then,

$$C_n(r, s; t, w) = C_n(1, \cdots, t, \cdots, w, \cdots, 1)$$

with the function on the right, $C_n(t_1, t_2, \cdots, t_n)$, the cycle indicator with $t_r = t$, $t_s = w$, and all other variables equal to unity. Then by equation (3*a*)

$$(1 - u) \exp uC(r, s; t, w) = \exp [(t - 1)(u^r/r) + (w - 1)(u^s/s)]$$

From this determine the recurrence relation

$$C_n(r, s; t, w) - nC_{n-1}(r, s, t, w) = \Sigma n!\, (t - 1)^j (w - 1)^k / r^j j!\, s^k k!$$

with $j$ and $k$ in the sum running over all pairs of integers such that $rj + sk = n$. Note that for $w = 1$, this is the same relation as in Problem 8.

14. Continuing Problem 13, derive the relation

$$\begin{aligned}(1 - u)&C(r, s; t, w) \exp uC(r, s; t, w) \\ &= [1 + (t - 1)(1 - u)u^{r-1} + (w - 1)(1 - u)u^{s-1}] \exp uC(r, s; t, w)\end{aligned}$$

with $C^n(r, s; t, w) \equiv C_n(r, s; t, w)$. From this, abbreviating $C_n(r, s; t, w)$ to $C_n$, find the recurrence relation

$$\begin{aligned}C_{n+1} = (n + 1)C_n + (t - 1)[(n)_{r-1}C_{n-r+1} - (n)_r C_{n-r}] \\ + (w - 1)[(n)_{s-1}C_{n-s+1} - (n)_s C_{n-s}]\end{aligned}$$

With $r = 1$, $s = 2$, verify the table

| $n$ | 0 | 1 | 2 | 3 | 4 |
|---|---|---|---|---|---|
| $C_n$ | 1 | $t$ | $t^2 + w$ | $2 + 3tw + t^3$ | $6 + 8t + 3w^2 + 6wt^2 + t^4$ |

and the recurrence $[C_n \equiv C_n(1, 2; t, w)]$

$$C_{n+1} = (n + t)C_n + (w - t)nC_{n-1} + (1 - w)n(n - 1)C_{n-2}$$

15. (*a*) With $C(n, r, s, j, k)$ the numbers of Problem 13, and the usual convention that $C^n(r, s, j, k) \equiv C(n, r, s, j, k)$, show that

$$(1 - u) \exp uC(r, s, j, k) = \frac{u^{rj+sk}}{r^j j!\, s^k k!} \exp (-u^r/r - u^s/s)$$

(*b*) Take $P(n, r, s)$ as the number of permutations in which the number of $r$-cycles is exactly the same as the number of $s$-cycles, so that

$$P(n, r, s) = \sum_j C(n, r, s, j, j)$$

Using the result above, show that, with $P^n(r, s) \equiv P(n, r, s)$, $i = \sqrt{-1}$,

$$(1 - u) \exp uP(r, s) = J_0(2iu^{(r+s)/2}/\sqrt{rs}) \exp(-u^r/r - u^s/s)$$

(Touchard, **8**)

16. (*a*) Write $d_n(t, r)$ for the enumerator of permutations of $n$ elements by number of cycles, none of which is an $r$-cycle. Show that with $d^n(t, r) \equiv d_n(t, r)$

$$\exp ud(t, r) = e^{-tu^r/r}(1 - u)^{-t}$$

$$d_{n+1}(t, r) = (n + t)d_n(t, r) - t(n)_{r-1}d_{n-r+1}(t) + t(n)_r d_{n-r}(t, r)$$

Note that $d_k(t, r) = t(t + 1) \cdots (t + k - 1)$, $\quad k < r, \quad d_n(t, 1) = d_n(t)$

(*b*) With $d_n(1, r) \equiv d_n(r)$, show that

$$d_n(r) = nd_{n-1}(r) + (-1)^k n!/k!\, r^k, \qquad n = kr$$
$$= nd_{n-1}(r), \qquad n \neq kr$$

Verify the following table:

| $n$ | 0 | 1 | 2 | 3 | 4 | 5 | 6 | 7 | 8 |
|---|---|---|---|---|---|---|---|---|---|
| $d_n(1)$ | 1 | 0 | 1 | 2 | 9 | 44 | 265 | 1854 | 14833 |
| $d_n(2)$ | 1 | 1 | 1 | 3 | 15 | 75 | 435 | 3045 | 24465 |
| $d_n(3)$ | 1 | 1 | 2 | 4 | 16 | 80 | 520 | 3640 | 29120 |
| $d_n(4)$ | 1 | 1 | 2 | 6 | 18 | 90 | 540 | 3780 | 31500 |

17. (*a*) A telephone exchange with $n$ subscribers is provided with means to connect subscribers in pairs only (no provision for conference circuits). The enumerator $T_n(t)$ for the number of connections is then given by

$$T_n(t) = C_n(1, t, 0, \cdots, 0)$$

with $C_n(t_1, t_2, \cdots, t_n)$ the cycle indicator, and, by equation (3*a*)

$$\exp uT(t) = \exp(u + tu^2/2), \qquad T^n(t) \equiv T_n(t)$$

Show that

$$T_{n+1}(t) = T_n(t) + ntT_{n-1}(t)$$

and verify the table

| $n$ | 0 | 1 | 2 | 3 | 4 | 5 |
|---|---|---|---|---|---|---|
| $T_n(t)$ | 1 | 1 | $1 + t$ | $1 + 3t$ | $1 + 6t + 3t^2$ | $1 + 10t + 15t^2$ |

(*b*) If $T_n(t) = \Sigma T(n, k)t^k$, show that

$$T(n, k) = n!/k!\,(n - 2k)!\,2^k$$
$$= T(n - 1, k) + (n - 1)T(n - 2, k - 1)$$

(*c*) If $U(n, k) = T(n, 0) + T(n, 1) + \cdots + T(n, k)$, show that

$$U(n, k) = U(n - 1, k) + (n - 1)U(n - 2, k - 1)$$

(*d*) Write $T_n \equiv T_n(1)$; derive the relations

$$T_n = T_{n-1} + (n - 1)T_{n-2} = U(n, m), \qquad m = [n/2]$$

$$= e^{-1/2} \sum_{k=m}^{\infty} \binom{2k}{n} n!/2^k k!, \qquad m = [(n + 1)/2]$$

and verify the table

| $n$ | 0 | 1 | 2 | 3 | 4 | 5 | 6 | 7 | 8 |
|---|---|---|---|---|---|---|---|---|---|
| $T_n$ | 1 | 1 | 2 | 4 | 10 | 26 | 76 | 232 | 764 |

*Note*: the Hermite polynomials $H_n(x)$ defined by $H_0(x) = 1$, $H_n(x) = 2xH_{n-1}(x) - 2(n-1)H_{n-2}(x)$ have the exponential generating function

$$\exp uH(x) = \exp(-u^2 + 2ux)$$

Hence, with $i = \sqrt{-1}$,

$$T_n(t) = \left(\frac{\sqrt{t}}{i\sqrt{2}}\right)^n H_n(i/\sqrt{2t})$$

and $T_n$ is asymptotically $(n/e)^{n/2}e^{\sqrt{n}}/e^{1/4}\sqrt{2}$ (compare Chowla et al., **2**).

18. (*a*) Take $e_n(t)$ as the enumerator of permutations of $n$ elements by number of even cycles, $o_n(t)$ the enumerator by number of odd cycles. (Note that these are *not* the same as the enumerators $c_n^e(t)$ and $c_n^o(t)$ of Section 6.) Show that

$$\exp ue(t) = (1 - u^2)^{-t/2}, \qquad e^n(t) \equiv e_n(t)$$
$$\exp uo(t) = (1 - u)^{-t/2}(1 + u)^{t/2}, \qquad o^n(t) = o_n(t)$$

(*b*) Derive the relations

$$(1 + u)^t \exp ue(t) = \exp uo(t)$$
$$(1 - u^2)e(t) \exp ue(t) = tu \exp ue(t)$$
$$(1 - u^2)o(t) \exp uo(t) = t \exp uo(t)$$

$$\sum_0^n \binom{n}{k} (t)_k e_{n-k}(t) = o_n(t)$$

$$e_{n+1}(t) = n(n - 1 + t)e_{n-1}(t)$$
$$o_{n+1}(t) = to_n(t) + n(n-1)o_{n-1}(t)$$

For verifications, the first few values are:

| $n$ | 0 | 1 | 2 | 3 | 4 | 5 |
|---|---|---|---|---|---|---|
| $e_n(t)$ | 1 | 0 | $t$ | 0 | $6t + 3t^2$ | 0 |
| $o_n(t)$ | 0 | $t$ | $t^2$ | $2t + t^3$ | $8t^2 + t^4$ | $24t + 20t^3 + t^5$ |

(*c*) Write $e_n = e_n(1)$, $o_n = o_n(1)$ for the total numbers; derive the results

$$o_n = e_n + ne_{n-1}$$
$$e_{n+1} = n^2 e_{n-1}$$
$$o_{n+1} = o_n + n(n-1)o_{n-1} = no_n + e_n$$
$$e_{2n} = ((2n)!/2^n n!)^2 = (2n-1)^2 e_{2n-2}$$
$$e_{2n+1} = 0$$
$$o_{2n} = (2n-1)o_{2n-1} = e_{2n}$$
$$o_{2n+1} = (2n+1)o_{2n}$$

Verify these results by the table

| $n$ | 0 | 1 | 2 | 3 | 4 | 5 | 6 | 7 | 8 |
|---|---|---|---|---|---|---|---|---|---|
| $e_n$ | 1 | 0 | 1 | 0 | 9 | 0 | 225 | 0 | 11025 |
| $o_n$ | 0 | 1 | 1 | 3 | 9 | 45 | 225 | 1575 | 11025 |

19. If $P$ is a permutation, if $P^k$ denotes $k$ iterations of $P$ considered as an operator (or as an element of the symmetric group), and 1 is the identity permutation, then for every $P$ there is an integer $m$, the order of $P$, such that $P^m = 1$ and $P^x \neq 1$, $x < m$. If $P$ is a cycle of length $k$, its order is $k$; if $P$ is composed of cycles of lengths $k_1, k_2, \cdots$, its order is the least common multiple of $k_1, k_2, \cdots$. If $T_n(m)$ is the number of permutations of $n$ elements satisfying $P^m = 1$, then each $P$ contains only cycles of length $m$ or one of its submultiples, and

$$\exp uT(m) = \exp \sum_{k|m} u^k/k, \qquad T^n(m) \equiv T_n(m)$$

$k|m$ indicating that the sum is over all divisors of $m$, including 1 and $m$ (compare Chowla et al., **3**). Show that

$$T_{n+1}(m) = \sum_{k|m} (n)_{k-1} T_{n-k+1}(m)$$

and, in particular, with $p$ a prime

$$T_{n+1}(p) = T_n(p) + (n)_{p-1} T_{n-p+1}(p)$$

Note that $T_n(p) = 1$, $n < p$, whereas $T_p(p) = 1 + (p-1)! \equiv 0 \pmod p$. Show that

$$T_n(p) \equiv 0, \qquad (\text{mod } p), \qquad n > p$$

$$T_{n+q}(p) \equiv T_n(p)T_q(p), \qquad (\text{mod } q)$$

$$T_{n+q}(m) \equiv T_n(m)T_q(m), \qquad (\text{mod } q)$$

(For the special case $p = m = 2$ of the last two, compare Moser and Wyman, **6**.)

20. (*a*) For the enumerators $c_n^e(t)$ and $c_n^o(t)$ of even and odd permutations by number of cycles, as in Section 6, show that

$$(1 - u^2)c^e(t) \exp uc^e(t) = t \exp uc^e(t) + tu \exp uc^o(t)$$

$$(1 - u^2)c^o(t) \exp uc^o(t) = tu \exp uc^e(t) + t \exp uc^o(t)$$

and, hence, that

$$c^e(t) \exp uc^e(t) - uc^o(t) \exp uc^o(t) = t \exp uc^e(t)$$

$$-uc^e(t) \exp uc^e(t) + c^o(t) \exp uc^o(t) = t \exp uc^o(t)$$

and

$$c_{n+1}^e(t) = tc_n^e(t) + nc_n^o(t)$$

$$c_{n+1}^o(t) = nc_n^e(t) + tc_n^o(t)$$

which are symmetrical in the two enumerators. Eliminating one of the enumerators, derive the "pure" recurrence relation

$$nx_{n+2} - (2n+1)tx_{n+1} + (n+1)(t^2 - n^2)x_n = 0$$

with $x_n = c_n^e(t)$ or $c_n^o(t)$. Verify these relations by the table

| $n$ | 0 | 1 | 2 | 3 | 4 | 5 |
|---|---|---|---|---|---|---|
| $c_n^e(t)$ | 1 | $t$ | $t^2$ | $2t + t^3$ | $11t^2 + t^4$ | $24t + 35t^3 + t^5$ |
| $c_n^o(t)$ | 0 | 0 | $t$ | $3t^2$ | $6t + 6t^3$ | $50t^2 + 10t^4$ |

(*b*) For the numbers $c_n^e = c_n^e(1)$, $c_n^o = c_n^o(1)$, determine the relations

$$(1 - u) \exp uc^e = 1 - u^2/2$$
$$(1 - u) \exp uc^o = u^2/2$$

and, hence, the values

$$c_0^e = c_1^e = 1, \qquad c_2^e = 1, \qquad c_n^e = nc_{n-1}^e = n!/2, \qquad n > 2$$
$$c_0^o = c_1^o = 0, \qquad c_2^o = 1, \qquad c_n^o = nc_{n-1}^o = n!/2, \qquad n > 2$$

which are verified by equations (39) and (40). $c_n^e$ is also the number of terms with positive sign of an $n$ by $n$ determinant, a result appearing in the solution to a problem by Pólya and Szegö.*

* G. Pólya and G. Szegö, *Aufgabe und Lehrsätze aus der Analysis*, vol. II, problem VII, 46, New York, 1945.

21. (*a*) Take $d_n^e(t)$, $d_n^o(t)$ as the enumerators for even and odd permutations by number of cycles, no one of which is a unit cycle. Derive the results:

$$2 \exp ud^e(t) = [(1 - u)^{-t} + (1 + u)^t] \exp(-tu)$$
$$2 \exp ud^o(t) = [(1 - u)^{-t} - (1 + u)^t] \exp(-tu)$$
$$d_{n+1}^e(t) = nd_n^o(t) + ntd_{n-1}^o(t)$$
$$d_{n+1}^o(t) = nd_n^e(t) + ntd_{n-1}^e(t)$$
$$x_{n+1} - n(n - 1)x_{n-1} - n(2n - 3)tx_{n-2} - n(n - 2)t^2x_{n-3} = 0$$

with $x_n = d_n^e(t)$ or $d_n^o(t)$.

Verify the table

| $n$ | 0 | 1 | 2 | 3 | 4 | 5 | 6 |
|---|---|---|---|---|---|---|---|
| $d_n^e(t)$ | 1 | 0 | 0 | $2t$ | $3t^2$ | $24t$ | $130t^2$ |
| $d_n^o(t)$ | 0 | 0 | $t$ | 0 | $6t$ | $20t^2$ | $120t + 15t^3$ |

(*b*) For the numbers $d_n^e \equiv d_n^e(1)$, $d_n^o \equiv d_n^o(1)$, derive the results:

$$(1 - u) \exp ud^e = (1 - u^2/2) \exp(-u)$$
$$(1 - u) \exp ud^o = (u^2/2) \exp(-u)$$
$$d_n^e = nd_{n-1}^e + (-1)^n \left[1 - \binom{n}{2}\right]$$
$$d_n^o = nd_{n-1}^o + (-1)^n \binom{n}{2}$$

and, with $D_n$ the displacement number of Chapter 3,

$$d_n^e + d_n^o = D_n$$
$$d_n^e - d_n^o = (-1)^{n-1}(n - 1)$$

In the notation of Pólya and Szegö (l.c. Problem 20), $D_n = \Sigma_n$ and $d_n^e - d_n^o = \Delta_n$; the determinant enumeration is for zero coefficients in the main diagonal.

22. Write $T_n^e(m)$ for the number of all even permutations of $n$ elements satisfying $P^m = 1$; then by equation (35)

$$2 \exp uT^e(m) = \exp \sum_{k|m} u^k/k + \exp \sum_{k|m} (-1)^{k-1}u^k/k$$

in the notation of Problem 19.

If

$$\exp uT^*(m) = \exp \sum_{k|m} (-1)^{k-1}u^k/k$$

then, with $T_n(m)$ as in Problem 19,

$$2T_n^e(m) = T_n(m) + T_n^*(m)$$

Hence for $m = p$, $p$ a prime

$$T_n^e(p) = T_n(p) = T_n^*(p)$$

Show that

$$T_{n+1}^*(m) = \sum (-1)^{k-1}(n)_{k-1}T_{n-k+1}^*(m)$$

$$T_{n+q}^*(m) \equiv T_n^*(m)T_q^*(m), \qquad (\text{mod } q)$$

$$T_{n+q}^e(m) \equiv T_n^e(m)T_q^e(m), \qquad (\text{mod } q)$$

CHAPTER 5

# Distributions: Occupancy

## 1. INTRODUCTION

Distribution has been defined by MacMahon, **1**, as the separation of a series of elements into a series of classes; more concretely, it may be described as the assignment of objects to boxes or cells. The objects may be of any number and kind and the cells may be specified in kind, capacity, and number independently. Order of objects in a cell may or may not be important. When the number of assignments is in question, the problem is said to be one of distribution; when the number of objects in given or arbitrary cells is in question, the problem is one of occupancy.

Problems of both kinds already have been noticed in Chapter 1, in relation to multinomial coefficients in Section 1.2.2, and in Examples 1.8 and 1.9. As suggested by these, the subject is closely related to permutations and combinations, but the point of view is sufficiently different to warrant a separate treatment.

## 2. UNLIKE OBJECTS AND CELLS

The simplest case is that of $n$ different objects and $m$ different cells. When there is no restriction on the occupancy of any cell, it follows at once from Example 1.11 that:

*The number of ways of putting n different objects into m different cells is* $m^n$.

More directly, it is clear that each object has a choice of $m$ cells; hence the number of choices for all $n$ is $m^n$, as stated.

This remark leads to a more extensive development as follows. Suppose that $x_i$ is the indicator of cell occupancy in the sense that $x_1^{n_1} x_2^{n_2} \cdots$ indicates a distribution of objects with $n_i$ in cell $i$. Then the indicator of distributions of a single object is

$$x_1 + x_2 + \cdots + x_m$$

and that of distributions of $n$ different objects is

$$(x_1 + x_2 + \cdots + x_m)^n$$

The exponential generating function for this is

$$\begin{aligned} F(t; x_1, x_2, \cdots, x_m) &= \sum_{n=0} (x_1 + x_2 + \cdots + x_m)^n t^n/n! \qquad (1) \\ &= \exp t(x_1 + x_2 + \cdots + x_m) \\ &= \exp tx_1 \exp tx_2 \cdots \exp tx_m \end{aligned}$$

Hence the enumerator for occupancy of cell $i$ is

$$\exp tx_i = 1 + x_i t + x_i^2 t^2/2! + \cdots + x_i^n t^n/n! + \cdots \qquad (2)$$

exactly as with permutations of objects of different kinds (Section 1.5). Thus expression (2) may be used for any desired specification of restrictions on occupancy in the familiar way. For example, if the cell may not be empty, the enumerator is $\exp tx_i - 1$; if the cell must contain at least $a$ and at most $a + b$, the enumerator is

$$x_i^a t^a/a! + x_i^{a+1} t^{a+1}/(a + 1)! + \cdots + x_i^{a+b} t^{a+b}/(a + b)!$$

Naturally the occupancy for each cell may be specified independently.

The generating function enumerating total possibilities is

$$F(t; 1, 1, \cdots, 1) = \exp mt$$

Equation (1) leads at once to another well-known result (which has appeared already in Example 1.12), namely:

*The number of ways of putting* n *different objects into* m *different cells, with no cell empty, is*

$$\Delta^m 0^n = m!\, S(n, m)$$

with $\Delta$ the finite difference operator: $\Delta u_n = u_{n+1} - u_n$, and $S(n, m)$ the Stirling number of the second kind.

In this case the enumerator for occupancy is

$$G(t; x_1, x_2, \cdots, x_m) = (\exp tx_1 - 1) \cdots (\exp tx_m - 1)$$

and

$$\begin{aligned} G(t; 1, 1, \cdots, 1) &= (e^t - 1)^m \\ &= \sum_{n=0} \Delta^m 0^n t^n/n! \end{aligned}$$

by Example 1.12 or Problem 2.14.

Further, if $p$ of the $m$ cells are occupied and the remaining cells are empty, the enumerator for total possibilities is

$$\binom{m}{p} (e^t - 1)^p = (m)_p \sum_{n=0} S(n, p) t^n/n!$$

since the $p$ occupied cells (or the $m - p$ empty cells) may be chosen in $\binom{m}{p}$ ways. Hence:

*The number of ways of putting* n *different objects into* m *different cells so that* p *cells are occupied and* m − p *are empty is*

$$(m)_p S(n, p)$$

Finally, it may be noticed that by the multinomial theorem

$$(x_1 + x_2 + \cdots + x_m)^n = \Sigma \frac{n!}{n_1! \cdots n_m!} x_1^{n_1} x_2^{n_2} \cdots x_m^{n_m}$$

so that

*The number of ways of putting $n$ different objects into $m$ different cells with $n_i$ in cell* i, i = 1, 2, · · ·, *m, is* $n!/n_1!\, n_2! \cdots n_m!$, which should be compared with Section 1.2.2.

Further examples and development appear in the problems.

## 3. LIKE OBJECTS AND UNLIKE CELLS

The propositions corresponding to those in the section ahead are as follows:

*The number of ways of putting $n$ like objects into $m$ different cells is*

$$\binom{n+m-1}{n} = \binom{n+m-1}{m-1}$$

*The same number when no cell is empty is*

$$\binom{n-1}{m-1}$$

Note that these results have appeared already in Examples 1.8 and 1.9.

They may be proved directly in the present setting as follows. First, for the case where no cell is empty, the number in question is the number of ways of placing $m - 1$ separators between $n$ like objects arranged in a row. There are $n - 1$ positions between the objects where separators may be placed, and the result follows at once. Next, when the restriction that no cell is empty is abandoned, the number is the same as that of $n + m$ like objects in $m$ cells with no cell empty, for the removal of $m$ of the $n + m$ objects, one from each cell, leaves the remaining $n$ distributed without restriction. Note also that

$$\sum_{k=0}^{m-1} \binom{m}{k} \binom{n-1}{m-k-1} = \sum_{k=1}^{m} \binom{m}{k} \binom{n-1}{k-1} = \binom{n+m-1}{n}$$

A proof which leads to a general development is as follows. The enumerator for occupancy of cell $i$ proper to like objects is obtained from equation (2) by removing the factorial denominators; hence it is

$$E(t; x_i) = 1 + x_i t + x_i^2 t^2 + \cdots + x_i^n t^n + \cdots = (1 - x_i t)^{-1} \tag{3}$$

The enumerator for all $m$ cells is

$$\begin{aligned} E(t; x_1, x_2, \cdots, x_m) &= E(t; x_1)E(t; x_2) \cdots E(t; x_m) \\ &= 1/(1 - x_1 t)(1 - x_2 t) \cdots (1 - x_m t) \\ &= 1/(1 - a_1 t + a_2 t^2 + \cdots + (-1)^m a_m t^m) \\ &= 1 + h_1 t + h_2 t^2 + \cdots + h_n t^n + \cdots \end{aligned} \tag{4}$$

with $a_k \equiv a_k(x_1, x_2, \cdots, x_m)$, the elementary symmetric function of weight $k$, and $h_k \equiv h_k(x_1, x_2, \cdots, x_m)$, the "homogeneous product sum" of weight $k$ [which may be regarded as defined by the last two equations of (4)].

Note that the enumerator for distribution of $n$ like objects in $m$ unlike cells is $h_n(x_1, x_2, \cdots, x_m)$ whereas that for $n$ unlike objects, by equation (1), is $(x_1 + x_2 + \cdots + x_m)^n$ or $h_1^n$ with $h_1 \equiv h_1(x_1, x_2, \cdots x_m) = x_1 + x_2 + \cdots + x_m$.

The enumerator for a given cell which may not be empty is

$$xt + x^2t^2 + \cdots = xt(1 - xt)^{-1}$$

whereas that for all cells, none of which may be empty, is

$$\begin{aligned} D(t; x_1, x_2, \cdots, x_m) &= x_1 x_2 \cdots x_m t^m/(1 - x_1 t) \cdots (1 - x_m t) \\ &= \sum_{k=0} a_m h_n t^{n+m} \end{aligned} \tag{5}$$

Then the enumerators for total possibilities are, respectively

$$E(t; 1, 1, \cdots, 1) = (1 - t)^{-m} = \Sigma \binom{n + m - 1}{n} t^n$$

$$D(t; 1, 1, \cdots, 1) = t^m(1 - t)^{-m} = \Sigma \binom{n - 1}{m - 1} t^n$$

in agreement with results stated above.

The distributions with a specified number of objects in each of the $m$ cells are given by $h_n(x_1, x_2, \cdots, x_m)$ in expanded form. Thus

$$h_1 = x_1 + x_2 + \cdots + x_m$$
$$h_2 = x_1^2 + x_2^2 + \cdots + x_m^2 + (x_1x_2 + x_1x_3 + \cdots + x_{m-1}x_m)$$

and it is clear that for any specification of the numbers $n_i$, $i = 1, 2, \cdots, m$, with $n_1 + n_2 + \cdots + n_m = n$, the answer is unity, as is otherwise evident.

Further examples and development, again, are to be found in the problems.

## 4. OBJECTS OF ANY SPECIFICATION AND UNLIKE CELLS

From the results now at hand, the distribution problem for objects of any kind and unlike cells can be solved completely. The general result with no restriction on the number of objects in any cell is as follows:

*The number of ways of putting n objects of specification* $(1^{n_1}2^{n_2}\cdots)$ *into m different cells is*

$$m^{n_1}\binom{m+1}{2}^{n_2}\binom{m+2}{3}^{n_3}\cdots$$

This follows from preceding results, in particular the first proposition of Section 3, and the fact that the objects of various kinds are distributed independently. Thus, for objects of specification $(pq)$, the $p$ of one kind may be distributed in $\binom{p+m-1}{p}$ ways, and the $q$ of the second kind are distributed in $\binom{q+m-1}{q}$ ways; by the rule of product, the number of ways of distributing objects of both kinds is

$$\binom{p+m-1}{p}\binom{q+m-1}{q}$$

To determine the number of ways when no cell may be empty, write $[n]$ for the partition $(1^{n_1}2^{n_2}\cdots)$ and

$$U([n], m) = m^{n_1}\binom{m+1}{2}^{n_2}\cdots \tag{6}$$

for the number of unrestricted distributions; write $R([n], m)$ for the number of distributions with no cell empty.

Then, classifying the distributions of $U([n], m)$ according to the number of empty cells, it follows at once that

$$\sum_{k=0}^{m}\binom{m}{k} R([n], m-k) = U([n], m) \tag{7}$$

or, what is the same thing,

$$U([n], m) = [R([n],) + 1]^m, \qquad R^k([n],) \equiv R([n], k)$$

The inverse of this is

$$R([n], m) = [U([n],) - 1]^m$$
$$= \sum_{k=0}^{m} (-1)^k \binom{m}{k} U([n], m - k) \qquad (8)$$

a result derived by a different method by P. A. MacMahon, **1**.

As (8) in fully developed form is unwieldy, it is helpful to notice some recurrences. Suppose first that some of the objects appear singly (the number $n_1$ in the specification is not zero) so that $[n] = (1^p s)$, with $s$ a partition without unit parts, and, for convenience, write $U([n], m)$ as $U_m(1^p s)$; then

$$U_m(1^p s) = mU_m(1^{p-1}s)$$

and (8) shows that

$$R_m(1^p s) = \Sigma\, (-1)^k \binom{m}{k} (m - k)U_{m-k}(1^{p-1}s)$$
$$= mR_m(1^{p-1}s) + mR_{m-1}(1^{p-1}s) \qquad (9)$$

In a similar way, if the specification contains 2-parts, as indicated by $(2^p s)$, with $s$ containing no 2-parts,

$$U_m(2^p s) = \binom{m+1}{2} U_m(2^{p-1}s)$$

and

$$R_m(2^p s) = \binom{m+1}{2} R_m(2^{p-1}s) + m^2 R_{m-1}(2^{p-1}s) + \binom{m}{2} R_{m-2}(2^{p-1}s) \quad (10)$$

The latter depends on the identity

$$\binom{m-k+1}{2} = \binom{m+1}{2} - mk + \binom{k}{2}$$

In the case where the specification contains $p$ $q$-parts,

$$U_m(q^p s) = \binom{m+q-1}{q} U_m(q^{p-1}s)$$
$$\binom{m-k+q-1}{q} = \Sigma(-1)^j \binom{m+q-j-1}{q-j} \binom{k}{j}$$

and

$$R_m(q^p s) = \sum_{j=0} \binom{m+q-j-1}{q-j} \binom{m}{j} R_{m-j}(q^{p-1}s) \qquad (11)$$

Another kind of recurrence follows from noting that

$$\binom{q+m}{q+1} = \frac{q+m}{q+1} \binom{q+m-1}{q}$$

so that

$$(q + 1)U_m((q + 1)s) = (q + m)U_m(qs)$$

and

$$(q + 1)R_m((q + 1)s) = (q + m)R_m(qs) + mR_{m-1}(qs) \tag{12}$$

The probability distribution for number of empty cells has the generating function

$$P(x; m, [n]) = \sum_{k=0} p(k; m, [n])x^k \tag{13}$$

with

$$p(k; m, [n]) = \binom{m}{k} R_{m-k}([n])/U_m[n]$$

The modifications of the enumerators, given by (4) and (5) for objects of a single kind [specification $(n)$] to account for general specification may be seen most easily in the next more general case, that of specification $(pq)$, that is, $p$ objects of one kind, $q$ of a second. If

$$E(t_1; x_1, x_2, \cdots, x_m) = 1 + h_1t_1 + \cdots + h_pt_1^p + \cdots$$

is the enumerator by number of objects for objects of the first kind and

$$E(t_2; x_1, x_2, \cdots, x_m) = 1 + h_2t_2 + \cdots + h_qt_2^q + \cdots$$

is the enumerator for objects of the second kind, then

$$E(t_1, t_2; x_1, x_2, \cdots, x_m) = \Sigma\Sigma h_ph_qt_1^pt_2^q \tag{14}$$

is the enumerator for objects of both kinds, since the objects are distributed independently. In (14) as in (4), the symmetric function $h_p$ is a function of the $m$ variables $x_1, x_2, \cdots, x_m$.

In the general case, as in (14), enumerating variables are introduced, one for each kind, and the general enumerator in all variables is the product of enumerators. The general result is as follows:

*The enumerator for occupancy of m different cells by objects of specification* $(1^{n_1}2^{n_2} \cdots)$ *is the symmetric function product*

$$h_1^{n_1}h_2^{n_2} \cdots$$

*with* $h_j = h_j(x_1, x_2, \cdots, x_m)$.

That this agrees with the proposition at the start of this section is seen at once when it is noticed that

$$h_j(1, 1, \cdots, 1) = \binom{m + j - 1}{j}$$

For occupancy problems, different enumerators are required. If as above $t_1, t_2, \cdots$ are the variables for objects of different kinds, the enumerator for single-cell occupancy is

$$\begin{aligned} 1/(1 - xt_1)(1 - xt_2) \cdots = 1 + xh_1(t_1, t_2, \cdots) + x^2h_2(t_1, t_2, \cdots) + \\ \cdots + x^nh_n(t_1, t_2, \cdots) + \cdots \end{aligned} \tag{15}$$

in which the term $(xt_1)^{n_1}(xt_2)^{n_2}\cdots$ indicates the occupancy of the given cell by $n = n_1 + n_2 + \cdots$ objects, $n_1$ of which are of the first kind, $n_2$ of the second, and so on. Of course (15) may be modified for restricted occupancy, with restrictions on the appearance of any kind of object, exactly as (2) and (3).

Since unlike cells are occupied independently, the enumerator for all $m$ cells is

$$O(x_1, x_2, \cdots, x_m; t_1, t_2, \cdots) = O(x_1; t_1, t_2, \cdots) \cdots O(x_m; t_1, t_2, \cdots) \tag{16}$$

with

$$\begin{aligned} O(x; t_1, t_2, \cdots) &= 1/(1 - t_1x)(1 - t_2x)\cdots \\ &= 1 + xh_1 + x^2h_2 + \cdots + x^nh_n + \cdots \end{aligned}$$

as given by (15).

When no cell is empty, and no attention is paid to which cell is occupied, the enumerator for occupancy is

$$(h_1x + h_2x^2 + \cdots)^m$$

This is the result from which MacMahon derived equation (8).

It should be noticed that both (15) and (16) put all object specifications in one basket, the $h_n$ symmetric functions, and thus force the consideration of all of these together. When particular sets of specifications such as $(21^{n-2})$, $(p1^{n-p})$ are in question, it is easier to proceed otherwise. Thus, for single cell occupancy, because of the independence of distribution of the several types, the probability distribution for any specification may be obtained as a simple product. Write

$$P(x; [n]) = \Sigma p(k, [n])x^k$$

for the probability generating function for single-cell occupancy by objects of specification $[n] = (1^{n_1}2^{n_2}\cdots)$. Then

$$P(x; [n] = [P(x; (1))]^{n_1}[P(x; (2))]^{n_2}\cdots \tag{17}$$

with

$$\begin{aligned} P(x; (1)) &= h_1(x, 1, \cdots, 1)/h_1(1, 1, \cdots, 1) = (x + m - 1)/m \\ P(x; (2)) &= h_2(x, 1, \cdots, 1)/h_2(1, 1, \cdots, 1) \\ &\cdot \\ &\cdot \\ &\cdot \\ P(x; (n)) &= h_n(x, 1, \cdots, 1)/h_n(1, 1, \cdots, 1) \end{aligned}$$

This follows from (4); note that $h_n$ is a function of $m$ variables $x_1, x_2, \cdots, x_m$, the indicators for cell occupancy.

Similar results may be written down for occupancy of any number of given cells, but, of course, with increasing complexity. They are illustrated in the problems.

The number of distributions with specified numbers of objects in each of the $m$ cells now requires development of the symmetric function product

$$h_1^{p_1}h_2^{p_2}\cdots$$

(for objects of specification $1^{p_1}2^{p_2}\cdots$) as a polynomial in its variables $x_1, x_2, \cdots, x_m$ or, more briefly, in terms of the monomial symmetric functions of these variables, that is, of functions like

$$(21) = \Sigma x_i^2 x_j$$
$$(2^2 1^2) = \Sigma x_i^2 x_j^2 x_k x_l$$

where the sum in the first is over all $i, j$ from 1 to $m$ with $i \neq j$, and, similarly, for the second. This development becomes complicated in short order and will not be entered upon here; the interested reader may consult MacMahon's treatise.

## 5. ORDERED OCCUPANCY

Here the objects in each cell are ordered, and distributions are counted as different when the cell ordering is different, even though the cell objects are alike. The problem is closely related to the distribution of like objects into unordered cells, as will be clear in the following results.

The first of these is:

*The number of ways of putting n different objects into m different ordered cells, with no restriction on the number in any cell is*

$$m(m+1)\cdots(m+n-1) = n!\binom{m+n-1}{n}$$

*If no cell may be empty, the number is*

$$n!\binom{n-1}{m-1}$$

This may be shown by first putting $n$ like objects into $m$ different cells without regard to cell ordering, then for each such distribution supposing the objects unlike and permuting them in all $n!$ ways, which accomplishes all possible orderings in the cells.

For like objects, all cell orderings are indifferent and the distributions are the same for ordered and unordered different cells. These suggest the general result which is as follows:

*The number of ways of putting objects of specification* $(n_1 n_2 \cdots)$ *into m different ordered cells is the product of*

$$n!/n_1!\,n_2!\cdots, \qquad n = n_1 + n_2 + \cdots$$

*and the corresponding number of ways of putting n like objects into m different unordered cells.*

The last number, of course, may include any occupancy restrictions desired, as discussed in Section 3 above.

Note that the relative occupancies, that is, the ratios of particularly specified occupancies to the total possibilities, are unchanged from those for $n$ like objects and $m$ different unordered cells.

## 6. LIKE CELLS

When cells are alike, the general problems of distribution and occupancy are of great involvement and only the simplest results are noticed here. First:

*The number of ways of putting n different things into m like cells, with no cells empty, is*

$$S(n, m)$$

*the Stirling number of the second kind, and the number without restriction is*

$$S(n, 1) + S(n, 2) + \cdots + S(n, m)$$

The first of these is obtained by dividing the corresponding number for unlike cells by $m!$, the number of ways of permuting the cells, since each of the distributions for like cells has $m!$ copies when the cells are made unlike. The second follows by simple summation, since empty cells may be chosen in just one way.

This is another combinatorial interpretation of the Stirling numbers $S(n, m)$ and it may be noticed that it is equivalent to the following: *A number which is the product of n different prime factors may be factored into* m *factors in* S(n, m) *ways.*

Also, when $m \geq n$, the numbers of distributions without restriction are the exponential numbers of Section 4.5 [see the remark following equation (4.25)].

Next, for like objects:

*The number of ways of putting n like things into m like cells, with no cells empty, is*

$$p_n(m)$$

*the number of partitions of n into m parts, and the number without restriction is*

$$p_n(1) + p_n(2) + \cdots + p_n(m)$$

The numbers $p_n(m)$ will be examined in detail in the next chapter. As is to be expected, these numbers have no simple general structure like that for $S(n, m)$.

As suggested by this result, the distributions for objects of general specification are equivalent to a more elaborate kind of partitioning, one in which each part is a composite number containing as many numbers as there are kinds of objects being distributed. Put in another way, each cell is described by a composite number describing the numbers of objects of each kind it contains. At least one of these numbers must be positive else the cell would be empty, and the sum of the numbers for any kind over all cells must be the given number of objects of this kind, else some object would be undistributed.

This is too large a subject to enter into here; the interested reader again is referred to MacMahon's treatise. Table 1 gives results for the simplest cases.

## REFERENCES


**1.** P. A. MacMahon, *Combinatory Analysis*, vol. I, Cambridge, 1915.
**2.** E. Netto, *Lehrbuch der Combinatorik*, Leipzig, 1901.
**3.** W. A. Whitworth, *Choice and Chance*, London, 1901.


## PROBLEMS

1. Write $D(n, m)$ for the number of ways of putting $n$ different things into $m$ different cells with no cell empty. From the relation

$$D(n, m) + mD(n, m-1) + \cdots + \binom{m}{k} D(n, m-k) + \cdots + D(n, 0) = m^n$$

or

$$[D(n, ) + 1]^m = m^n = E^m 0^n, \qquad D(n, )^k \equiv D(n, k)$$

show that

$$D(n, m) = \sum_{k=0}^{m} (-1)^k \binom{m}{k} (m-k)^n = \Delta^m 0^n$$

in agreement with the text.

2. For $n$ different objects distributed into $m$ different cells, the probability of $s$ empty cells is

$$p(s; n, m) = (m)_{m-s} S(n, m-s)/m^n$$

Show that, with $\Delta$ the usual finite difference operator,

$$\begin{aligned} P(x; n, m) &= \Sigma p(s; n, m) x^s \\ &= (\Delta + x)^m 0^n/m^n \\ M(x; n, m) &= P(1 + x; n, m) \\ &= (E + x)^m 0^n/m^n, \qquad E^k 0^n \equiv k^n \end{aligned}$$

and, hence, that the binomial moments $B_k(n, m)$ are given by

$$B_k(n, m) = \binom{m}{k}\left(1 - \frac{k}{m}\right)^n$$

In particular

$$B_1(n, m) = m\left(1 - \frac{1}{m}\right)^n = \text{expected number of empty cells}$$

$$m - B_1(n, m) = m\left[1 - \left(1 - \frac{1}{m}\right)^n\right] = \text{expected number of occupied cells}$$

3. (*a*) Take $p(s; n, m, k)$ as the probability that exactly $s$ cells each have exactly $k$ objects when $n$ unlike objects are distributed into $m$ unlike cells. Show that

$$\sum_{n=0} m^n p(s; n, m, k)t^n/n! = \binom{m}{s}\left(\frac{t^k}{k!}\right)^s \left(e^t - \frac{t^k}{k!}\right)^{m-s}$$

and hence that

$$\sum_{n=0} P(x; n, m, k)t^n/n! = \left(e^{t/m} + (x - 1)\frac{(t/m)^k}{k!}\right)^m$$

with

$$P(x; n, m, k) = \sum_{s=0} p(s; n, m, k)x^s$$

(*b*) Show that the binomial moments of the probability distribution just derived are given by

$$B_j(n, m, k) = \binom{m}{j}\frac{n!}{k!^j(n - kj)!}\frac{(m - j)^{n-kj}}{m^n}$$

4. Show that the enumerator for occupancy of any cell when $n$ distinct objects are distributed over $m$ distinct cells is

$$(x + m - 1)^n$$

Hence, if $p(k; m, n)$ is the probability that the given cell contains $k$ objects

$$\begin{aligned} P(x; m, n) &= \Sigma p(k; m, n)x^k \\ &= (x + m - 1)^n/m^n \end{aligned}$$

and

$$p(k; m, n) = \binom{n}{k}\left(1 - \frac{1}{m}\right)^{n-k} m^{-k}$$

Show that binomial moments are given by

$$B_k(m, n) = \binom{n}{k} m^{-k}$$

and that the mean and variance are $n/m$ and $(n/m)(1 - 1/m)$.

5. Show similarly that the probability generating function for occupancy of a given pair of cells is

$$\begin{aligned} P(x, y) &= (x + y + m - 2)^n/m^n \\ &= \Sigma p(j, k; m, n)x^j y^k \end{aligned}$$

so that

$$p(j, k; m, n) = \binom{n}{j+k} (m-2)^{n-j-k} \binom{j+k}{j} / m^n$$

The binomial moment generating function is

$$B(x, y) = P(x+1, y+1) = \Sigma B_{jk}(m, n) x^j y^k$$

Show that

$$B_{jk}(m, n) = \binom{n}{j+k} \binom{j+k}{j} m^{-j-k}$$

and that the covariance is $-n/m^2$.

6. If each of $m$ distinct cells may contain $0, 1, 2, \cdots, s$ of $n$ distinct objects, the enumerator for the total number of ways of putting objects into cells is

$$\begin{aligned} f(t; m, s) &= \Sigma f_n(m, s) t^n/n! \\ &= (1 + t + t^2/2! + \cdots + t^s/s!)^m \end{aligned}$$

(a) Derive the recurrences

$$f_n(m, s) = f_n(m-1, s) + n f_{n-1}(m-1, s) + \cdots + \binom{n}{s} f_{n-s}(m-1, s)$$

$$f_{n+1}(m, s) = m f_n(m, s) - m \binom{n}{s} f_{n-s}(m-1, s)$$

(b) Using Problem 2.20, show that with $s > 1$

$$\begin{aligned} f_n(m, s) &= m^n \quad n \leq s \\ f_{s+1}(m, s) &= m^{s+1} - m \\ f_{s+2}(m, s) &= m^{s+2} - m - (m)_2(s+2) \\ f_{s+3}(m, s) &= m^{s+3} - m - (m)_2 \binom{s+4}{2} - (m)_3 \binom{s+3}{2} \end{aligned}$$

7. Continuing Problem 6, if each cell must contain at least $s$ objects, the enumerator for total number of ways is

$$\begin{aligned} g(t; m, s) &= (t^s/s! + t^{s+1}/(s+1)! + \cdots)^m \\ &= (e^t - 1 - t - \cdots - t^{s-1}/(s-1)!)^m \\ &= \Sigma g_n(m, s) t^n/n! \end{aligned}$$

Derive the recurrences

$$g_n(m, s) = m g_{n-1}(m, s) + m \binom{n-1}{s-1} g_{n-s}(m-1, s)$$

$$g_n(m, s) = \sum_{k=0}^{m} (-1)^k \binom{m}{k} \frac{n!}{(s-1)!^k (n-sk+k)!} g_{n-sk+k}(m-k, s-1)$$

8. (a) Write $L(n, m)$ for the number of ways of putting $n$ like objects into $m$ different cells with no cells empty. From the relation

$$L(n, m) + mL(n, m-1) + \cdots + \binom{m}{k} L(n, m-k) + \cdots + L(n, 0) = \binom{n+m-1}{n}$$

show that

$$L(n, m) = \sum_{k=0} (-1)^k \binom{m}{k} \binom{n+m-k-1}{n}$$

(*b*) Prove the identity

$$\binom{n-1}{m-1} = \sum_{k=0} (-1)^k \binom{m}{k} \binom{n+m-k-1}{n}$$

*Hint*: use the generating function identity

$$t^m(1-t)^{-m} = \Sigma(-1)^k \binom{m}{k} (1-t)^{-m+k}$$

9. If $p(k; n, m)$ is the probability of putting $n$ like objects into $m$ different cells leaving $k$ cells empty, show that

$$\begin{aligned} p(k; n, m) &= \binom{m}{k} L(n, m-k) \Big/ \binom{n+m-1}{n} \\ &= N(k; n, m) \Big/ \binom{n+m-1}{n} \end{aligned}$$

with the latter a definition.

Derive the relations

$$\sum_{n=0} t^n \sum_{k=0} x^k N(k; n, m) = (x - 1 + (1-t)^{-1})^m$$

$$\begin{aligned} \Sigma t^n \Sigma (1+x)^k N(k; n, m) &= (x + (1-t)^{-1})^m \\ &= \Sigma t^n \, \Sigma x^k \binom{m}{k} \binom{n+m-k-1}{n} \end{aligned}$$

and, from these, find the binomial moments of the probability distribution $p(k; n, m)$, $k = 0, 1, \cdots, m$ to be

$$B_k(n, m) = \binom{m}{k} \binom{n+m-k-1}{n} \Big/ \binom{n+m-1}{n}$$

In particular, show that the mean and variance are

$$m(m-1)(n+m-1)^{-1} \quad \text{and} \quad m(m-1)n(n-1)/(n+m-1)^2(n+m-2)$$

10. In a similar way, if $N(k; n, m, s)$ is the number of ways of distributing $n$ like objects into $m$ different cells with exactly $k$ cells having exactly $s$ objects, show that

$$[(x-1)t^s + (1-t)^{-1}]^m = \Sigma t^n \Sigma x^k N(k; n, m, s)$$

and that the binomial moments of the distribution $p(k; n, m, s)$, $k = 0, 1, \cdots, m$ with

$$p(k; n, m, s) = N(k; n, m, s) \Big/ \binom{n+m-1}{n}$$

are

$$B_k(n, m, s) = \binom{m}{k} \binom{n+m-k(s+1)-1}{n-ks} \Big/ \binom{n+m-1}{n}$$

11. Show that the enumerator for occupancy of a single cell when $n$ like objects are put into $m$ different cells is

$$E(t, x; m) = (1 - xt)^{-1}(1 - t)^{-m+1}$$
$$= \Sigma t^n \Sigma x^j \binom{n + m - j - 2}{n - j}$$
$$= \Sigma t^n \Sigma \binom{n + m - 1}{n - j} (x - 1)^j$$

and, hence, that the binomial moments of the corresponding distribution are given by

$$B_j(n, m) = \binom{n + m - 1}{n - j} \Big/ \binom{n + m - 1}{n} = (n)_j/(m + j - 1)_j$$

The mean and variance are $n/m$ and $n(n + m)(m - 1)/(m + 1)m^2$ respectively.

12. Show similarly that the binomial moments for joint occupancy of two cells are given by

$$B_{jk}(n, m) = (n)_{j+k}/(m + j + k - 1)_{j+k}$$

and, in particular, that the covariance is $-n(n + m)/(m + 1)m^2$.

13. (*a*) For restricted occupancy when each of $m$ different cells may contain no more than $p$ of $n$ like objects, so that the enumerator of total number of ways is

$$E(t; m, p) = (1 + t + \cdots + t^p)^m = \Sigma t^n N_n(m, p)$$

show that

$$N_n(m, p) = \sum_{k=0}^{m} (-1)^k \binom{m}{k} \binom{n - kp - k + m - 1}{n - kp - k}$$

(*b*) Derive the recurrences

$$N_n(m, p) = N_n(m - 1, p) + N_{n-1}(m - 1, p) + \cdots + N_{n-p}(m - 1, p)$$
$$N_n(m, p) = N_{n-1}(m, p) + N_n(m - 1, p) - N_{n-p-1}(m - 1, p)$$

14. Continuing Problem 13, show that the binomial moments of the distribution by number of empty cells are given by

$$B_j(n, m, p) = \binom{m}{j} \frac{N_n(m - j, p)}{N_n(m, p)}$$

with $N_n(m, p)$ as defined in Problem 13.

15. Show similarly that the binomial moments of the distribution by number of cells containing exactly $k$ objects are

$$B_j(k, n, m, p) = \binom{m}{j} \frac{N_{n-jk}(m - j, p)}{N_n(m, p)}$$

16. For restricted occupancy when each of $m$ different cells may contain no less than $p$ of $n$ like objects, show that the enumerator for total number of distributions is

$$E(t; m, p) = (t^p + t^{p+1} + \cdots)^m = t^{mp}(1 - t)^{-m}$$
$$= \sum_{n=0} \binom{n + m - 1}{n} t^{n+mp}$$
$$= \sum_{n=mp} \binom{n - mp + m - 1}{m - 1} t^n$$

17. For restricted occupancy, as in Problem 16, and with the additional restriction that no cell contain more than $p + q - 1$ objects, verify the results

$$E(t; m, p, q) = t^{pm}(1 - t^q)^m(1 - t)^{-m}$$
$$= \sum_{n=mp} t^n N_n(m, p, q)$$

with

$$N_{n+mp}(m, p, q) = \sum_{k=0}^{m} (-1)^k \binom{m}{k} \binom{n - kq + m - 1}{m - 1}$$

Compare with Problem 13($a$).

18. For objects of specification $(p1^{n-p})$ and $U_m(p1^{n-p})$, the number of unrestricted distributions into $m$ different cells show that

$$\sum_{p=0} \sum_{n-p=0} U_m(p1^{n-p}) \frac{t^p u^{n-p}}{(n - p)!} = e^{um}(1 - t)^{-m}$$

and that

$$\Sigma\Sigma R_m(p1^{n-p}) \frac{t^p u^{n-p}}{(n - p)!} = \left(\frac{e^u}{1 - t} - 1\right)^m$$

with $R_m(p1^{n-p})$ the corresponding number with no cell empty.

Derive the result

$$R_m(p1^q) = \sum_{k=0}^{m} \binom{m}{k} \binom{p + k - 1}{m - 1} \Delta^k 0^q$$

19. For objects of specification $(pq)$, $m$ different cells, and $R_m(pq)$, as in Problem 18 and the text, use equation (12) of the text (with appropriate change of notation) to show that

$$R_m(pq) = \sum_{k=0} \binom{p - 1}{k} \binom{q}{m - k - 1} \binom{m + p - 1 - k}{p}$$

20. For objects of specification $(2^p s)$, $m$ different cells, and $R_m(2^p s)$ defined as usual, if

$$R(x; 2^p s) = \sum_{m=1} R_m(2^p s) x^{m-1} = R_p(x)$$

show that

$$R_p(x) = (1 + 4x + 3x^2)R_{p-1}(x) + (2x + 5x^2 + 3x^3)R'_{p-1}(x)$$
$$+ (1/2)x^2(1 + x)^2 R''_{p-1}(x)$$

with primes denoting derivatives.

21. For occupancy of one of $m$ different cells by objects of specification $(pq)$, show that binomial moments are given by

$$B_k = \sum_{j=0}^{k} \binom{p+m-1}{p-j}\binom{q+m-1}{q-k+j} \Big/ \binom{p+m-1}{p}\binom{q+m-1}{q}$$

22. For occupancy of one of $m$ different cells by objects of specification $(1^{p_1}2^{p_2}\cdots)$, show that the generating function of binomial moments is given by

$$B(x; m, 1^{p_1}2^{p_2}\cdots) = B(x; m, 1)^{p_1} B(x; m, 2)^{p_2}\cdots$$

with

$$B(x; m, n) = \sum_{j=0} x^j \binom{n+m-1}{n-j} \Big/ \binom{n+m-1}{n}$$

23. Write $A_n(m)$ for the number of ways of putting $n$ different things into $m$ like cells so that

$$A_n(m) = S(n, 1) + S(n, 2) + \cdots + S(n, m)$$
$$= A_n(m-1) + S(n, m)$$

Show that

$$A_n(m) = \sum_{i=0}^{m} d_{m-i} i^n / i!$$
$$= \frac{1}{m!}(D+E)^m 0^n, \qquad D^k \equiv D_k, \qquad E^k 0^n = k^n$$

with $D_n = n!\, d_n = \Delta^n 0! =$ rencontres number.

TABLE 1

DISTRIBUTION OF OBJECTS INTO LIKE CELLS, NO CELL EMPTY

| No. Cells | Object Specification | | | | | | |
|---|---|---|---|---|---|---|---|
| | (2) | $(1^2)$ | | | | | |
| 1 | 1 | 1 | | | | | |
| 2 | 1 | 1 | | | | | |
| | (3) | (21) | $(1^3)$ | | | | |
| 1 | 1 | 1 | 1 | | | | |
| 2 | 1 | 2 | 3 | | | | |
| 3 | 1 | 1 | 1 | | | | |
| | (4) | (31) | $(2^2)$ | $(21^2)$ | $(1^4)$ | | |
| 1 | 1 | 1 | 1 | 1 | 1 | | |
| 2 | 2 | 3 | 4 | 5 | 7 | | |
| 3 | 1 | 2 | 3 | 4 | 6 | | |
| 4 | 1 | 1 | 1 | 1 | 1 | | |
| | (5) | (41) | (32) | $(31^2)$ | $(2^21)$ | $(21^3)$ | $(1^5)$ |
| 1 | 1 | 1 | 1 | 1 | 1 | 1 | 1 |
| 2 | 2 | 4 | 5 | 7 | 8 | 11 | 15 |
| 3 | 2 | 4 | 6 | 8 | 11 | 16 | 25 |
| 4 | 1 | 2 | 3 | 4 | 5 | 7 | 10 |
| 5 | 1 | 1 | 1 | 1 | 1 | 1 | 1 |

CHAPTER 6

# Partitions, Compositions, Trees, and Networks

## 1. INTRODUCTION

According to L. E. Dickson, **4**, to whose account of the history the reader is referred for many interesting results, partitions first appeared in a letter from Leibniz to Johann Bernoulli (1669). The real development starts, like so much else in combinatoric, with Euler (1674).

The use of partitions in specifying a collection of objects of various kinds has already appeared in Chapter 1, and naturally calls for an enumeration. A partition by definition is a collection of integers (with given sum) without regard to order. It is natural, therefore, to consider along with partitions the corresponding ordered collections, which, following P. A. MacMahon, are called compositions. Trees and networks belong in the same frame of reference because of similarity of generating function, as will appear.

The integers collected to form a partition (or composition) are called its parts, and the number which is the sum of these parts is the partitioned (or composed) number. It is conventional to abbreviate repeated parts by the use of exponents; for example, partition 111 is written $1^3$. Table 1 shows all partitions of $n$ for $n = 1(1)8$ grouped by number of parts. Partitions and compositions may be considered for enumeration with or without restrictions on the kind, size, number of repetitions, or number of parts. The unrestricted partitions and compositions of numbers 3 and 4 are

| Number | Partitions | Compositions |
|---|---|---|
| 3 | 3, 21, $1^3$ | 3, 21, 12, $1^3$ |
| 4 | 4, 31, $2^2$, $21^2$, $1^4$ | 4, 31, 13, $2^2$, 211, 121, 112, $1^4$ |

TABLE 1

PARTITIONS OF $n$ BY NUMBER OF PARTS

| $n$ | Number of Parts 1 | 2 | 3 | 4 | 5 | 6 | 7 | 8 |
|---|---|---|---|---|---|---|---|---|
| 1 | 1 | | | | | | | |
| 2 | 2 | $1^2$ | | | | | | |
| 3 | 3 | 21 | $1^3$ | | | | | |
| 4 | 4 | 31 | $21^2$ | $1^4$ | | | | |
| | | $2^2$ | | | | | | |
| 5 | 5 | 41 | $31^2$ | $21^3$ | $1^5$ | | | |
| | | 32 | $2^21$ | | | | | |
| 6 | 6 | 51 | $41^2$ | $31^3$ | $21^4$ | $1^6$ | | |
| | | 42 | 321 | $2^21^2$ | | | | |
| | | $3^2$ | $2^3$ | | | | | |
| 7 | 7 | 61 | $51^2$ | $41^3$ | $31^4$ | $21^5$ | $1^7$ | |
| | | 52 | 421 | $321^2$ | $2^21^3$ | | | |
| | | 43 | $3^21$ | $2^31$ | | | | |
| | | | $32^2$ | | | | | |
| 8 | 8 | 71 | $61^2$ | $51^3$ | $41^4$ | $31^5$ | $21^6$ | $1^8$ |
| | | 62 | 521 | $421^2$ | $321^3$ | $2^21^4$ | | |
| | | 53 | 431 | $3^21^2$ | $2^31^2$ | | | |
| | | $4^2$ | $42^2$ | $32^21$ | | | | |
| | | | $3^22$ | $2^4$ | | | | |

Partitions may also be exhibited by an array of dots called the Ferrers graph; for example, the graph for 532 is

. . . . .
. . .
. .

As shown, there is one row for each part, the number of dots in each row indicates the part size, and all rows are aligned on the left.

The partition obtained by reading a Ferrers graph by columns is called the conjugate (of the given partition); the conjugate of 532 is 33211, and vice versa. Several enumeration identities are simple consequences of conjugacy, as will appear below.

Compositions may also be exhibited by an array of dots called a zigzag graph by MacMahon, its inventor, which is like the Ferrers graph except that the first dot of each part is aligned with the last dot of its predecessor; thus the graph for composition 532 is

```
. . . . .
        . . .
            . .
```

As with partitions, the conjugate composition is obtained by reading the graph by columns; the conjugate of 532 is 11112121.

A tree is a special form of network or linear graph (the term "graph" is used both for the Ferrers and zigzag diagrams mentioned above and for networks because this is standard terminology; no confusion should arise since the meaning is assured by the context). A linear graph is defined informally as a collection of points, and a collection of lines or

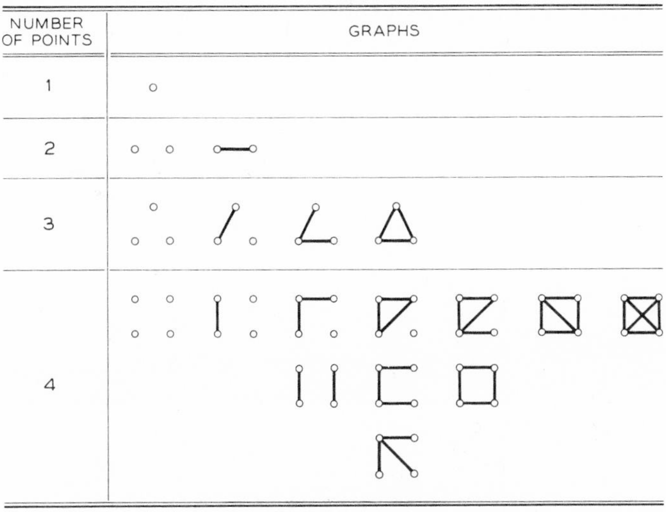

NUMBER OF POINTS | GRAPHS

1

2

3

4

**Fig. 1.** Linear graphs.

pairs of points which describes the connections of points; a collection of $n$ discrete points (without connecting lines) is a graph and, at the other extreme, a collection of $n$ points, each pair of which is joined by a line [so that there are $n(n-1)/2$ lines in total], is a graph, the complete graph. For enumeration purposes it is convenient to follow the restrictions that just one line may join two points (no lines in parallel, in electrical terminology) and that the graph has no "slings", lines joining points to themselves, although in the general theory of graphs (see König, **15**) neither of these restrictions is made. The distinct linear graphs with $n$ points, $n = 1(1)4$, are shown in Fig. 1.

A *cycle* of a graph is a collection of lines of the form: $p_1p_2, p_2p_3, \cdots, p_{k-1}p_1$, where $p_ip_j$ designates the line joining points $p_i$ and $p_j$, and all points in the collection save $p_1$ are distinct. A graph is *connected* if every pair of points is joined by a *path*, that is, a collection of lines of the form $p_1p_2, p_2p_3, \cdots, p_{k-1}p_k$, with all points $p_1$ to $p_k$ distinct.

A connected linear graph without cycles (or lines in parallel or "slings") is a *tree*. This mathematical object has a closer affinity to a family tree than to the growing varieties. A tree with one point, the root, distinguished from all other points by this very fact, is called a *rooted* tree and, to emphasize the contrast, an unrooted tree is called a *free* tree.

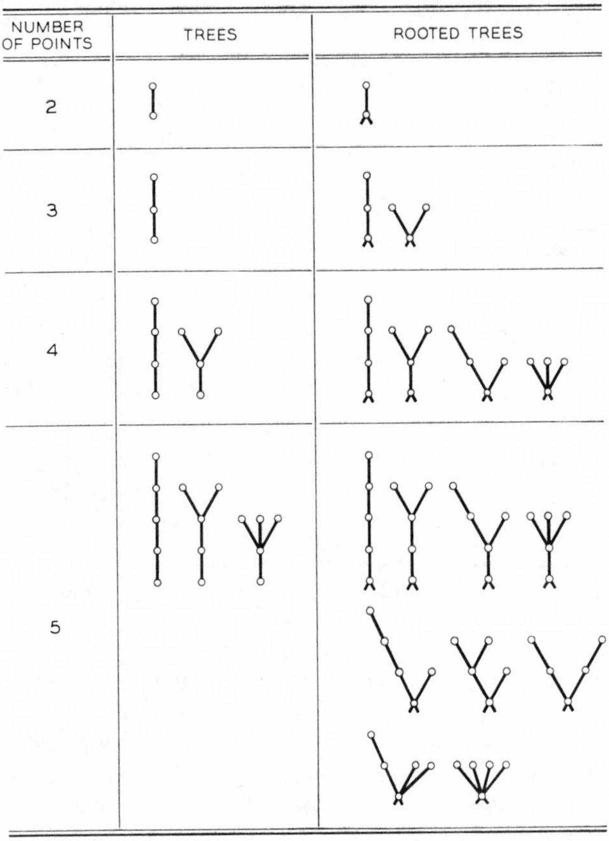

| NUMBER OF POINTS | TREES | ROOTED TREES |
|---|---|---|
| 2 | | |
| 3 | | |
| 4 | | |
| 5 | | |

**Fig. 2.** Trees and rooted trees with all points alike.

Fig. 2 shows all distinct trees and rooted trees with $n$ points, $n = 2$ to 5. Because of the absence of cycles, a tree has a number of points one greater than the number of its lines. The rooted trees may be regarded as obtained from trees by making each point in turn the root and eliminating duplicates.

The enumerations of trees and rooted trees in the first instance are by number of points, the points tacitly being assumed alike. Then points and lines are given added characteristics, the points being *labeled* or *colored*, the lines *labeled, colored*, or *oriented*. These terms will be explained later. The enumerations of linear graphs in the first instance are by numbers of points and of lines, but, again, either points or lines or both may be given added characteristics, and, of course, subgraphs other than trees may also be examined. The possible enumerations are of enormous number and the material given in the text and the problems merely broaches the subject.

## 2. GENERATING FUNCTIONS FOR PARTITIONS

Write $p_n$ for the number of unrestricted partitions of $n$ and, with $p_0 = 1$,

$$p(t) = p_0 + p_1 t + p_2 t^2 + \cdots$$

for its generating function. Then

**Theorem 1.** *The enumerating generating function for unrestricted partitions is*

$$\begin{aligned} p(t) &= (1 + t + t^2 + t^3 + \cdots)(1 + t^2 + t^4 + \cdots) \cdots \\ &\qquad (1 + t^k + t^{2k} + t^{3k} + \cdots) \cdots \qquad (1) \\ &= 1/(1 - t)(1 - t^2) \cdots (1 - t^k) \cdots \end{aligned}$$

Proof is immediate by noticing that each factor of the product accounts for all possible contributions of parts of a given size. Notice that (1) is also the generating function for combinations with repetition in which the first element is unrestricted, the second has appearances which are multiples of two, the third multiples of three, and so on.

Generating functions for certain kinds of restricted partitions may be written down from (1) at sight. Thus, the partitions with no part greater than $k$ are enumerated by

$$\begin{aligned} p_k(t) &= \Sigma p_{nk} t^n \\ &= 1/(1 - t)(1 - t^2) \cdots (1 - t^k) \end{aligned} \qquad (2)$$

The partitions with no repeated parts or, what is the same thing, with unequal parts are enumerated by

$$u(t) = (1 + t)(1 + t^2)(1 + t^3) \cdots \qquad (3)$$

and those with every part odd are enumerated by

$$o(t) = 1/(1 - t)(1 - t^3)(1 - t^5) \cdots \qquad (4)$$

Since $(1 - t^{2k}) = (1 - t^k)(1 + t^k)$, it follows that

$$u(t)(1 - t)(1 - t^2)(1 - t^3) \cdots = (1 - t^2)(1 - t^4)(1 - t^6) \cdots$$

or

$$u(t) = o(t) \tag{5}$$

which may be stated as

**Theorem 2.** *The partitions with unequal parts are equinumerous with those with all parts odd.*

For example, the partitions of 5 with unequal parts are 5, 41, 32, whereas those with odd parts are 5, $31^2$, $1^5$.

For enumerating by number of parts, a two-variable generating function, also due to Euler, is necessary. If

$$\begin{aligned} F(t, u) &= (1 + ut + u^2t^2 + u^3t^3 + \cdots)(1 + ut^2 + u^2t^4 + \cdots) \cdots \\ &\quad (1 + ut^k + u^2t^{2k} + \cdots) \cdots \\ &= 1/(1 - ut)(1 - ut^2) \cdots (1 - ut^k) \cdots \end{aligned} \tag{6}$$

and

$$F(t, u) = \Sigma p(t, k)u^k \tag{7}$$

then $p(t, 0) = 1$ and $p(t, k)$ is the generating function for partitions with exactly $k$ parts. This is so because the term $u^j t^{jk}$ in the general factor of (6) is the indicator for $j$ appearances of part $k$.

To find $p(t, k)$, note that

$$(1 - ut)F(t, u) = F(t, tu) \tag{8}$$

or, using (7),

$$p(t, k) - tp(t, k - 1) = t^k p(t, k) \tag{9}$$

Hence

$$(1 - t^k)p(t, k) = tp(t, k - 1)$$

$$(1 - t^{k-1})(1 - t^k)p(t, k) = t^2 p(t, k - 2)$$

and so on, until finally, since $p(t, 0) = 1$,

$$p(t, k) = t^k/(1 - t)(1 - t^2) \cdots (1 - t^k) \tag{10}$$

If the partitions enumerated by this function are labeled $p_n(k)$, where, of course, $p_n(k)$ is the number of partitions of $n$ with exactly $k$ parts, comparison of (10) and (2) gives at once

$$p_{nk} = p_{n+k}(k) \tag{11}$$

Otherwise stated, this result is

**Theorem 3.** *The number of partitions of n with no part greater than k equals the number of partitions of n + k with exactly k parts.*

For example, the partitions of 6 with no part greater than 2 are $2^3$, $2^21^2$, $21^4$, $1^6$, four in number, whereas those of 8 with exactly two parts are 71, 62, 53, 44.

It follows from $p(t, 0) = 1$, and mathematical induction using (10), that

$$\begin{aligned} P(t, k) &= p(t,0) + p(t, 1) + \cdots + p(t, k) \\ &= 1/(1-t)(1-t^2)\cdots(1-t^k) \\ &= t^{-k}p(t, k) = p_k(t) \end{aligned} \tag{12}$$

Hence

**Theorem 3(a).** *The number of partitions of n with at most k parts equals the number of partitions of n with no part greater than k, and also the number of partitions of n + k with exactly k parts.*

For $n = 6$, the partitions with at most 2 parts are 6, 51, 43, $3^2$, four in number as required by the theorem.

A bridge between Theorems 3 and 3($a$) is made by considering conjugate partitions. The conjugate of a partition with $k$ parts is a partition with one or more parts equal to $k$; hence, the first half of Theorem 3($a$).

The function for enumerating partitions with unequal parts by number of parts is

$$\begin{aligned} G(t, a) &= (1 + at)(1 + at^2)(1 + at^3)\cdots \\ &= \Sigma u(t, k)a^k \end{aligned} \tag{13}$$

with $u(t, k)$ being the generating function for partitions with $k$ unequal parts. Then

$$G(t, a) = (1 + at)G(t, at) \tag{14}$$

and, equating coefficients,

$$u(t, k) = t^k u(t, k) + t^k u(t, k-1) \tag{15}$$

Hence, by iterations of (15) and $u(t, 0) = 1$

$$u(t, k) = t^{\binom{k+1}{2}}/(1-t)(1-t^2)\cdots(1-t^k) \tag{16}$$

This result may be stated as

**Theorem 3(b).** *The partitions described in Theorem 3(a) are equal in number to the partitions of* $n + \binom{k+1}{2}$ *with k unequal parts.*

The four partitions of 9 ($n = 6, k = 2$) with 2 unequal parts are 81, 72, 63, 54.

## 3. USE OF THE FERRERS GRAPH

As has been noticed above, reading the Ferrers graph by rows and by columns gives an enumeration identity which may be put as

**Theorem 4.** *The number of partitions of $n$ with exactly $m$ parts equals the number of partitions into parts the largest of which is $m$.*

Thus the 3-part partitions of 6 are $41^2$, $321$, $2^3$, whereas those with largest part 3 are $31^3$, $321$, $3^2$. The first part of Theorem 3(*a*) is a direct consequence of Theorem 4, as has been noticed.

Other uses of the graph are equally instructive. Take one of the equalities of Theorem 3(*h*) in the following form: *the number of partitions of $n$ with exactly $k$ parts equals the number of partitions of $n + \binom{k}{2}$ with $k$ unequal parts.* For orientation, take $n = 6$, $k = 3$, so that $n + \binom{k}{2} = 9$; the partitions in question are $41^2$, $321$, $2^3$ and $621$, $531$, $432$. Looking at the graph of 621, namely,

```
· · · · · ·
· ·
·
```

it is clear that it may be separated into

```
· · · ·    and    · ·
·                 ·
·
```

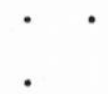

the first being $41^2$, one of the 3-part partitions of 6.

The same is true for 531 and 432, in each case involving the triangle shown above, that is, the partition (210).

It is clear that, for arbitrary $k$, the addition to any $k$-rowed graph with $n$ dots of a triangle graph with $k - 1$ rows aligned as above will produce a graph of $n + \binom{k}{2}$ dots with no two rows alike, which is what is required by the theorem.

Consider now self-conjugate partitions which read the same by rows and columns, for example, 321 with graph

```
· · ·
· ·
·
```

Every such graph has a symmetry which may be expressed as follows: (i) there is a corner square of $m^2$ dots (called the Durfee square), and when this is removed (ii) there are two like "tails", which represent partitions of $(n - m^2)/2$ into at most $m$ parts. Given $m^2$ and $n$, there is a class of self-conjugate partitions the number of which is the coefficient of $t^{(n-m^2)/2}$ in

$$p_m(t) = 1/(1 - t)(1 - t^2) \cdots (1 - t^m)$$

or, putting $t = x^2$, the coefficient of $x^n$, in

$$x^{m^2}/(1 - x^2)(1 - x^4) \cdots (1 - x^{2m})$$

The generating function for self-conjugate partitions is then

$$1 + \frac{x}{1 - x^2} + \frac{x^4}{(1 - x^2)(1 - x^4)} + \frac{x^9}{(1 - x^2)(1 - x^4)(1 - x^6)} + \cdots \quad (17)$$

Now read the self-conjugate graph by angles as in

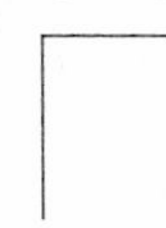

Since rows and columns are equal, each angle contains an odd number of dots and for any given partition no two angles are alike. Hence the total number of such graphs is also the number of partitions with unequal odd parts, enumerated by

$$(1 + x)(1 + x^3)(1 + x^5) \cdots$$

Moreover, if the corner square has $m^2$ dots, reading by angles results in $m$ parts, and the enumerator by number of parts is

$$(1 + ax)(1 + ax^3)(1 + ax^5) \cdots$$

The identity arrived at finally is

$$(1 + ax)(1 + ax^3)(1 + ax^5) \cdots$$
$$= 1 + a\frac{x}{1 - x^2} + a^2\frac{x^4}{(1 - x^2)(1 - x^4)} + a^3\frac{x^9}{(1 - x^2)(1 - x^4)(1 - x^6)} + \cdots \quad (18)$$

A final illustration is a beautiful proof due to Fabian Franklin* of a famous Euler identity concerning the coefficients in the expansion of

$$(1 - t)(1 - t^2)(1 - t^3) \cdots$$

This is equation (13) with $a = -1$; hence the coefficient of $t^n$ is generated by $u(t, 0) + u(t, 2) + u(t, 4) + \cdots - [u(t, 1) + u(t, 3) + \cdots]$. This

* *C.R. Acad. Sci. Paris*, vol. 92 (1881), pp. 448–50.

coefficient may be written $E(n) - O(n)$ where $E(n)$ is the number of partitions of $n$ with an even number of unequal parts, $O(n)$ the number with an odd number of unequal parts; for example $E(7) = 3$, the partitions being 61, 42, 43, and $O(7) = 2$, partitions 7 and 421.

Franklin's idea was to establish a correspondence between $E(n)$ and $O(n)$ which would isolate the difference.

Suppose a partition into unequal parts has a graph like the following (for 76532)

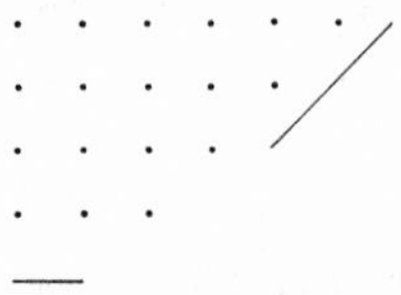

```
. . . . . . /
. . . . . /
. . . . /
. . .
___
```

with a base line (which may be a single point) at the bottom, and an outer diagonal (45 degrees, which is certain to include no interior points) which may also be a single point. Suppose the number of points in the base is $b$, and the number in the diagonal is $d$. Moving the base alongside the diagonal reduces the number of parts by one; moving the diagonal below the base increases the number of parts by one, in either case changing the parity (evenness or oddness) of the number of parts. When are one or the other of these movements possible without changing the character of the graph (unequal parts in descending order)?

There are three cases: $b < d$, $b = d$, and $b > d$. For $b < d$, the base may be moved. For $b = d$, the base again may be moved (but not the diagonal) unless the base meets the diagonal as in

```
. . . . .
. . . .
. . .
```

For $b > d$, the diagonal may be moved unless, again, the two lines meet and $b = d + 1$. In the first exceptional case (for $b = d$), the partition is of the form

$$b, b + 1, b + 2, \cdots, 2b - 1$$

and

$$n = b + (b + 1) + (b + 2) + \cdots + (2b - 1) = (3b^2 - b)/2$$

In the second exceptional case the partition is

$$d + 1, d + 2, \cdots, 2d$$

and

$$n = (3d^2 + d)/2$$

Hence $E(n) - O(n)$ is zero unless $n = (3k^2 \pm k)/2$, when $E(n) - O(n) = (-1)^k$, and the Euler identity may be written as

$$\begin{aligned}(1-t)(1-t^2)(1-t^3)\cdots \\ &= 1 - t - t^2 + t^5 + t^7 - t^{12} - t^{15} + \cdots \qquad (19) \\ &= 1 + \sum_1^\infty (-1)^k (t^{(3k^2-k)/2} + t^{(3k^2+k)/2})\end{aligned}$$

This identity is especially interesting in association with (1) since

$$p(t)(1-t)(1-t^2)(1-t^3)\cdots = 1$$

and, equating coefficients,

$$p_n - p_{n-1} - p_{n-2} + p_{n-5} + p_{n-7} - \cdots + (-1)^k[p_{n-k_1} + p_{n-k_2}] + \cdots = 0 \qquad (20)$$

with $k_1 = (3k^2 - k)/2$, $k_2 = (3k^2 + k)/2$. Equation (20) is a recurrence relation for unrestricted partitions, actually used by MacMahon to calculate $p_n$ up to $n = 200$.

## 4. DENUMERANTS

The term "denumerant" was introduced by Sylvester to denote the number of partitions into specified parts, repeated or not; thus $D(n; a_1, a_2, \cdots, a_m)$ denotes the number of partitions of $n$ into parts $a_1, a_2, \cdots, a_m$ or, what is the same thing, the number of solutions in integers of

$$a_1x_1 + a_2x_2 + \cdots + a_mx_m = n$$

The corresponding generating function is

$$\begin{aligned}D(t; a_1, a_2, \cdots, a_m) &= \Sigma D(n; a_1, a_2, \cdots, a_m)t^n \\ &= 1/(1-t^{a_1})(1-t^{a_2})\cdots(1-t^{a_m})\end{aligned}$$

In this notation, the number of unrestricted partitions is the denumerant $D(n; 1, 2, 3, \cdots)$, the number with no part greater than $k$ the denumerant $D(n; 1, 2, 3, \cdots, k)$, the number with all parts odd $D(n; 1, 3, 5, \cdots)$, and so on.

The simplest mode of evaluation is that used by Euler, namely, evaluation by recurrence. Thus, since

$$(1-t^k)D(t; 1, 2, 3, \cdots, k) = D(t; 1, 2, 3, \cdots, k-1)$$

it follows at once that

$$\begin{aligned}D(n; 1, 2, 3, \cdots, k) - D(n-k; 1, 2, 3, \cdots, k) \\ = D(n; 1, 2, 3, \cdots, k-1) \qquad (21)\end{aligned}$$

Since $D(n; 1) = 1$, $n > 0$ (single partition $1^n$), this serves as a step-by step evaluation for every $k$ and, of course,

$$D(n; 1, 2, 3, \cdots, k) = D(n; 1, 2, 3, \cdots), \qquad n < k$$

so the unrestricted partitions are also obtained. However, an error at any stage can never be surely compensated.

Cayley avoided this inescapable defect of recurrence by partial-fraction expansions; for example,

$$\frac{1}{(1-t)(1-t^2)} = \frac{1}{2(1-t)^2} + \frac{1}{4(1-t)} + \frac{1}{4(1+t)}$$

$$\frac{1}{(1-t)(1-t^2)(1-t^3)} = \frac{1}{6(1-t)^3} + \frac{1}{4(1-t)^2} + \frac{17}{72(1-t)} + \frac{1}{8(1+t)} + \frac{2+t}{9(1+t+t^2)}$$

From these, expressions for denumerants are found, like

$$4D(n; 12) = 2n + 3 + (1, -1) \text{ pcr } 2_n \tag{22}$$

$$72D(n; 123) = 6n^2 + 36n + 47 + 9(1, -1) \text{ pcr } 2_n + 8(2, -1, -1) \text{ pcr } 3_n \tag{23}$$

These are written in a notation introduced by Cayley, **2**; pcr is read prime circulant and $(1, -1)$ pcr $2_n$ is shorthand for a function which has the value 1 for $n$ even and $-1$ for $n$ odd.

It will be noticed that the partial-fraction expansion has been carried out completely for the repeated factor $(1 - t)$, that the quadratic $1 + t + t^2$ has not been factored, and that the presence of a repeated factor has complicated the procedure.

For the denumerant $D(n; 1, 2, 3, \cdots, k)$, a simpler expansion has been given by MacMahon. This proceeds from the remark that the symmetric functions $h_i$ of the variables $\alpha_1, \alpha_2, \cdots$ have the generating function (compare Problem 2.27($b$)):

$$1 + h_1x + h_2x^2 + \cdots = 1/(1 - \alpha_1x)(1 - \alpha_2x)(1 - \alpha_3x) \cdots$$

If the variables $\alpha_1, \alpha_2, \cdots$ are taken as $1, t, t^2, \cdots$, then

$$1 + h_1x + h_2x^2 + \cdots = 1/(1 - x)(1 - tx)(1 - t^2x) \cdots$$

and

$$h_k = P(t, k) = 1/(1 - t)(1 - t^2) \cdots (1 - t^k) \tag{24}$$

On the other hand, the same substitution of variables carries the power sum symmetric functions $s_1, s_2, \cdots$ defined by

$$s_i = \alpha_1^i + \alpha_2^i + \cdots$$

into

$$s_i = 1 + t^i + t^{2i} + \cdots = 1/(1 - t^i) \tag{25}$$

Hence the relation between these two kinds of symmetric functions is a partial-fraction expansion for denumerant generating functions. As has already appeared in Problem 2.27,

$$1 + h_1x + h_2x^2 + \cdots = \exp(s_1x + s_2x^2/2 + s_3x^3/3 + \cdots)$$

or

$$\begin{aligned} k!h_k &= Y_k(s_1, s_2, 2s_3, \cdots, (k-1)!s_k) \\ &= C_k(s_1, s_2, s_3, \cdots, s_k) \end{aligned}$$

where $Y_k$ is a Bell polynomial and $C_k$ is the cycle indicator of Chapter 4.

The first instances of this relation, namely,

$$2h_2 = s^2 + s_2$$

$$6h_3 = s^3 + 3s_1s_2 + 2s_3$$

correspond to

$$\frac{2}{(1-t)(1-t^2)} = \frac{1}{(1-t)^2} + \frac{1}{1-t^2}$$

$$\begin{aligned} \frac{6}{(1-t)(1-t^2)(1-t^3)} &= \frac{1}{(1-t)^3} + \frac{3}{(1-t)(1-t^2)} + \frac{2}{1-t^3} \\ &= \frac{1}{(1-t)^3} + \frac{3}{2(1-t)^2} + \frac{3}{2(1-t^2)} + \frac{2}{1-t^3} \end{aligned}$$

(the last expression by use of the first). From these it is found that

$$2D(n; 12) = n + 1 + (1, 0) \text{ pcr } 2_n$$

$$12D(n; 123) = (n+1)(n+5) + 3(1, 0) \text{ pcr } 2_n + 4(1, 0, 0) \text{ pcr } 3_n$$

both of which are simpler than their Cayley correspondents, equations (22) and (23). Noting that $(n + 1)(n + 5) = n^2 + 6n + 5$ and that the circulant terms contribute 0, 3, 4 or 7 leads to De Morgan's result that $D(n; 123)$ is the nearest integer to $(n + 3)^2/12$.

In either type of partial-fraction expansion, the circulant terms remain as a nuisance. They may be avoided entirely, as noticed by E. T. Bell, **3**, if the variable $n$ of the denumerant is properly chosen. For $D(n; 123)$ notice that

$$D(6n; 123) = 3n^2 + 3n + 1 \qquad D(6n + 3; 123) = 3n^2 + 6n + 3$$
$$D(6n + 1; 123) = 3n^2 + 4n + 1 \qquad D(6n + 4; 123) = 3n^2 + 7n + 4$$
$$D(6n + 2; 123) = 3n^2 + 5n + 2 \qquad D(6n + 5; 123) = 3n^2 + 8n + 5$$

and that 6 is the least common multiple of the parts 1, 2, 3.

The general expression of this result, which has been proved by E. T. Bell, **3**, is

**Theorem 5.** *If $a$ is the least common multiple of $a_1, a_2, \cdots, a_m$, the denumerant $D(an + b; a_1, a_2, \cdots, a_m)$, $b = 0, 1, 2, \cdots, a - 1$, is a polynomial in $n$ of degree $m - 1$, that is to say,*

$$D(an + b) = c_0 + c_1 n + \cdots + c_m n^{m-1}$$

*where $c_0, c_1, \cdots, c_m$ are constants independent of $n$.*

The constants are fully determined when the denumerant is known for $m$ different values of $n$, say $n_1, n_2, \cdots, n_m$, or, what is the same thing, $D(an + b)$ may be expressed uniquely in terms of $D(an_1 + b)$, $D(an_2 + b)$, $\cdots$, $D(an_m + b)$. In fact, by Lagrange's interpolation formula

$$D(an + b) = \sum_{j=1}^{m} \frac{F_j(n)}{F_j(n_j)} D(an_j + b) \tag{26}$$

where $F(x) = (x - n_1)(x - n_2) \cdots (x - n_m)$, and $F_j(x) = F(x)/(x - n_j)$. Putting $n_j = j$, this becomes

$$D(an + b) = \binom{n-1}{m} \sum_{j=1}^{m} (-1)^{m-j} \binom{m}{j} \frac{jD(aj + b)}{n - j} \tag{27}$$

Thus the denumerant is determined for all values of $n$ when it is known for $am$ values of $n$.

For three parts $a_1, a_2, a_3$, equation (27) becomes

$$2D(an + b) = (n - 2)(n - 3)D(a + b) - 2(n - 1)(n - 3)D(2a + b) + (n - 1)(n - 2)D(3a + b)$$

Finally, it should be noted that Gupta, **10**, whose tables of (unrestricted) partitions are the most extensive, used a procedure differing from any yet described. This is as follows. Let $r(n, m)$ be the number of partitions of $n$ with one part equal to $m$ and all other parts equal or greater than $m$; its enumerator is

$$r_m(t) = \Sigma r(n, m)t^n = \frac{t^m}{(1 - t^m)(1 - t^{m+1})(1 - t^{m+2})} \cdots \tag{28}$$

and hence

$$r(n, m) = D(n - m; m, m + 1, \cdots)$$

Then

$$p(n) = r(n, 1) + r(n, 2) + \cdots + r(n, n) \tag{29}$$

since the partitions enumerated by $r(n, j)$ differ from those enumerated by $r(n, k)$, for $k$ unequal to $j$. Alternatively (29) follows from summing on $m$ from 1 to infinity the identity

$$\frac{1}{(1 - t^m)(1 - t^{m+1})} \cdots - \frac{t^m}{(1 - t^m)(1 - t^{m+1})} \cdots = \frac{1}{(1 - t^{m+1})(1 - t^{m+2})} \cdots$$

or

$$t^{-m}r_m(t) = r_m(t) + t^{-m-1}r_{m+1}(t)$$

Summing the same identity from $m$ to infinity, and to $m + k$, shows that

$$t^{-m}r_m(t) = r_m(t) + r_{m+1}(t) + \cdots \tag{30}$$

$$t^{-m}r_m(t) = r_m(t) + r_{m+1}(t) + \cdots + r_{m+k}(t) + t^{-m-k-1}r_{m+k+1}(t) \tag{30$a$}$$

From (30) it follows that

$$r(n, m) = r(n - m, m) + r(n - m, m + 1) + \cdots + r(n - m, n - m) \tag{31}$$

The corresponding result coming from (30$a$) is

$$r(n, m) = r(n - m, m) + r(n - m, m + 1) + \cdots + r(n - m, m + k) + r(n - k - 1, m + k + 1) \tag{32}$$

and in particular ($k = 0$)

$$r(n, m) = r(n - m, m) + r(n - 1, m + 1) \tag{33}$$

These equations, along with the evident boundary relations $r(k, m) = 0$, $k < m$, $r(m, m) = 1$, are sufficient for calculating all partitions by recurrence. The calculations are much abbreviated by noting the following results due to Gupta, **10**:

$$r(m + j, m) = 0, \qquad 0 < j < m$$

$$r(2m + j, m) = \sum_0^j r(m + j, m + i) = 1, \qquad 0 \leq j < m$$

$$\begin{aligned} r(3m + j, m) &= \sum_0^a r(2m + j, m + i) + r(3m + j + a + 1, m + a + 1) \\ &= 2 + a, \qquad a = [j/2], \qquad 0 \leq j < m \end{aligned}$$

and, with $b = [j/3]$,.

$$r(4m + j, m) = \sum_0^b (3m + j, m + i) + r(4m + j + b + 1, m + b + 1)$$
$$= 3 + [(m + j)/2] + [j(j + 6)/12]$$

Also

$$r(4m + k, > m) = r(4m + k, m + 1) + r(4m + k, m + 2) + \cdots$$
$$= m + 2 + [k/2] + [(m^2 + 2km + k^2 - 9)/12]$$
$$k < 4$$

Table 2 gives the unrestricted partitions of $n$ for the range 1 to 99.

TABLE 2

THE NUMBER OF UNRESTRICTED PARTITIONS OF $n$

$n = 10k + m$

| $m\backslash k$ | 0 | 1 | 2 | 3 | 4 | 5 | 6 |
|---|---|---|---|---|---|---|---|
| 0 | | 42 | 627 | 5604 | 37338 | 2 04226 | 9 66467 |
| 1 | 1 | 56 | 792 | 6842 | 44583 | 2 39943 | 11 21505 |
| 2 | 2 | 77 | 1002 | 8349 | 53174 | 2 81589 | 13 00156 |
| 3 | 3 | 101 | 1255 | 10143 | 63261 | 3 29931 | 15 05499 |
| 4 | 5 | 135 | 1575 | 12310 | 75175 | 3 86155 | 17 41630 |
| 5 | 7 | 176 | 1958 | 14883 | 89134 | 4 51276 | 20 12558 |
| 6 | 11 | 231 | 2436 | 17977 | 1 05558 | 5 26823 | 23 23520 |
| 7 | 15 | 297 | 3010 | 21637 | 1 24754 | 6 14154 | 26 79689 |
| 8 | 22 | 385 | 3718 | 26015 | 1 47273 | 7 15220 | 30 87735 |
| 9 | 30 | 490 | 4565 | 31185 | 1 73525 | 8 31820 | 35 54345 |

| $m\backslash k$ | 7 | 8 | 9 |
|---|---|---|---|
| 0 | 40 87968 | 157 96476 | 566 34173 |
| 1 | 46 97205 | 180 04327 | 641 12359 |
| 2 | 53 92783 | 205 06255 | 725 33807 |
| 3 | 61 85689 | 233 38469 | 820 10177 |
| 4 | 70 89500 | 265 43660 | 926 69720 |
| 5 | 81 18264 | 301 67357 | 1046 51419 |
| 6 | 92 89091 | 342 62962 | 1181 14304 |
| 7 | 106 19863 | 388 87673 | 1332 30930 |
| 8 | 121 32164 | 441 08109 | 1501 98136 |
| 9 | 138 48650 | 499 95925 | 1692 29875 |

It is worth noting that the Gupta procedure is related to the Euler recurrence relation of equation (20) as follows. First

$$r_1(t) = \frac{t}{(1 - t)(1 - t^2)(1 - t^3)} \cdots = tp(t)$$

hence, $r(n, 1) = p(n - 1)$. Next

$$r_2(t) = \frac{t^2}{(1 - t^2)(1 - t^3)} \cdots = t^2(1 - t)p(t)$$

and $r(n, 2) = p(n - 2) - p(n - 3)$. In the same way

$$r(n, 3) = p(n - 3) - p(n - 4) - p(n - 5) + p(n - 6)$$
$$r(n, 4) = p(n - 4) - p(n - 5) - p(n - 6) + p(n - 8) + p(n - 9) - p(n - 10)$$

Hence equation (29) corresponds to

$$\begin{aligned} p(n) = p(n - 1) &+ [p(n - 2) - p(n - 3)] + [p(n - 3) - p(n - 5) \\ &+ p(n - 6)] + [p(n - 4) - p(n - 5) - p(n - 6) \\ &+ p(n - 8) + p(n - 9) - p(n - 10)] + \cdots \end{aligned}$$

which shows the gradual build-up of equation (20).

## 5. PERFECT PARTITIONS

A perfect partition of a number $n$ is one which contains just one partition of every number less than $n$ when repeated parts are regarded as indistinguishable. Thus, $1^n$ is a perfect partition for every $n$; and for $n = 7$, $41^3$, $421$, $2^31$ and $1^7$ are all perfect partitions.

If the parts of a partition are regarded as weights for a weighing scale, then perfect partitions are solutions of the problem of determining a set of weights which, put together in one scale pan, will weigh any object of integral weight in just one way (or establish the weight of any object to within one weight unit in just one way). A similar interpretation relates perfect partitions to resistance boxes for electrical measurements. Finally they have a natural relation to number systems with a given base.

To determine all perfect partitions, notice first that there must be at least one unit part (the lone partition of 1). Suppose there are $q_1 - 1$ unit parts; then all numbers less than $q_1$ have just one partition, and a part $q_1$ must be taken as the next part. With $q_2 - 1$ added parts $q_1$, all numbers up to $q_1q_2$ are uniquely expressible with parts 1 and $q_1$. Continuing in this way, a perfect partition may be written

$$1^{q_1-1}q_1^{q_2-1}(q_1q_2)^{q_3-1}(q_1q_2q_3)^{q_4-1} \cdots (q_1q_2 \cdots q_{k-1})^{q_k-1}$$

Hence,

$$\begin{aligned} n &= q_1 - 1 + q_1(q_2 - 1) + \cdots + (q_1q_2 \cdots q_{k-1})(q_k - 1) \\ &= q_1q_2 \cdots q_k - 1 \end{aligned}$$

The numbers $q_1, q_2, \cdots, q_k$ *in this order* completely specify the partition, and $q_i > 1$ for every $i$ so,

**Theorem 6.** *The number of perfect partitions of n is the same as the number or ordered factorizations of n + 1 without unit factors.*

The factorizations of 8 are 8, 42, 24, and 222 which correspond to the perfect partitions $1^7$, $1^34$, $12^3$, and 124 mentioned above.

## 6. COMPOSITIONS

The enumeration of compositions is closely allied to that of combinations with repetition. If $c_m(t)$ is the enumerating generating function for compositions with exactly $m$ parts, then

$$\begin{aligned} c_m(t) &= \Sigma c_{m,n} t^n \\ &= (t + t^2 + t^3 + \cdots)^m = t^m(1 - t)^{-m} \end{aligned} \tag{34}$$

and

$$c_{m,n} = \binom{n-1}{m-1}$$

Note that there are $m$ like factors in the generating function indicating equal treatment of all parts and positions in a composition, and that each factor may be regarded as accounting for a particular position of the composition, and so contains terms for all possible part sizes.

The compositions with exactly $m$ parts, no one of which is greater than $s$, have the generating function

$$c_m(t; s) = (t + t^2 + \cdots t^s)^m \tag{35}$$

The compositions with exactly $m$ odd parts have the generating function

$$o_m(t) = (t + t^3 + t^5 + \cdots)^m \tag{36}$$

and the generating function for compositions with no (odd) part greater than $2s + 1$ is

$$o_m(t; s) = (t + t^3 + t^5 + \cdots + t^{2s+1})^m \tag{37}$$

The generating function with no restriction on the number or size of parts is

$$c(t) = \sum_{m=1}^{\infty} c_m(t) = \frac{t}{1 - 2t} \tag{38}$$

and

$$c_n = 2^{n-1} \tag{39}$$

Similarly, the generating function without restriction on the number of parts but with no part greater than $s$ is

$$c(t; s) = \sum_{m=1}^{\infty} c_m(t; s) = \frac{t - t^{s+1}}{1 - 2t + t^{s+1}} \tag{40}$$

Other generating functions are written down with similar ease and are left to the problems. But notice as a final example that, if the first $k$ ($k < m$) parts are specified to be $s_1 s_2 \cdots s_k$ *in this order*, the generating function for such compositions with $m$ parts is

$$t^{s_1+s_2+\cdots+s_k}(t + t^2 + \cdots)^{m-k}$$

If the positions of the specified parts are not specified, the generating function is

$$\binom{m}{k} t^{s_1+s_2+\cdots+s_k}(t + t^2 + \cdots)^{m-k}$$

If, further, no other parts are to be equal to the specified parts, the generating function is

$$t^{s_1+s_2+\cdots+s_k}(t + t^2 + \cdots - t^{s_1} - t^{s_2} - \cdots - t^{s_k})^{m-k}$$

## 7. ENUMERATION OF ROOTED TREES

This is the simplest of tree enumerations and is examined first for orientation. Both cases, where the points other than the root are all alike or all unlike, are treated. The distinct points may be regarded as kept distinct by labels or numbers, and it is convenient to label the root also, though, of course, this is not necessary. Figure 3 shows all distinct rooted trees with $n$ distinct points for $n = 1$ to 5 (for $n = 4, 5$, the figure shows only the number of labelings for each rooted tree with like points); as already mentioned, the rooted trees with like points are also shown in Fig. 2.

For the first case, take $r_n$ as the number of distinct rooted trees with $n$ points, all points other than the root being alike. Take $r_n(m)$ as the corresponding number when $m$ lines meet at the root, so that

$$r_n = \sum_{m=1} r_n(m)$$

Then, first $r_n(1) = r_{n-1}$, since these rooted trees may be formed by adding a line at the root to rooted trees with $n - 1$ points and then shifting the root. In the same way, for $n$ even,

$$r_{2q}(2) = r_1 r_{2q-2} + r_2 r_{2q-3} + \cdots + r_{q-1} r_q$$

but for $n$ odd

$$r_{2q+1}(2) = r_1 r_{2q-1} + r_2 r_{2q-2} + \cdots + r_{q-1} r_{q+1} + \binom{r_q + 1}{2}$$

since, when rooted trees with the same number of points are added to the two lines at the root, the number of possibilities, because of symmetry of

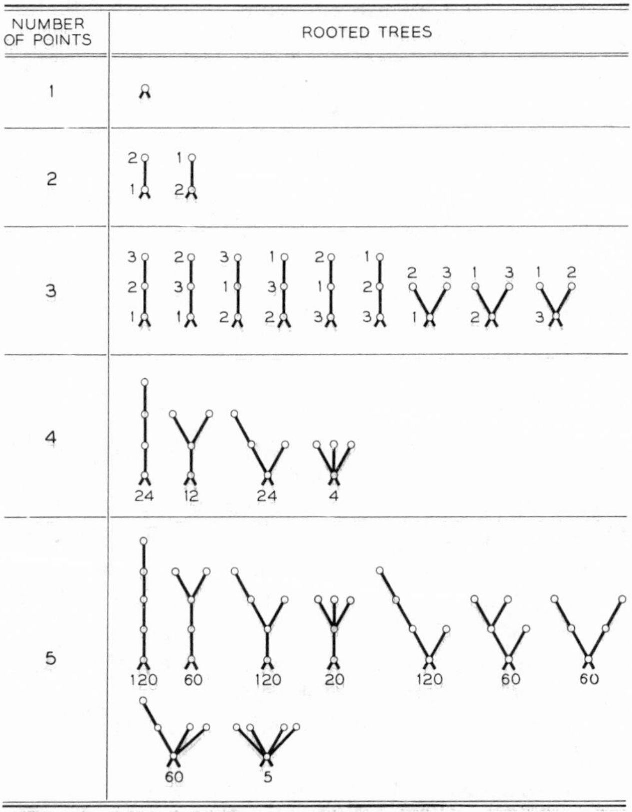

| NUMBER OF POINTS | ROOTED TREES |
|---|---|
| 1 | |
| 2 | |
| 3 | |
| 4 | 24 12 24 4 |
| 5 | 120 60 120 20 120 60 60 60 5 |

Fig. 3. Rooted trees with all points labeled.

the lines at the root, is not the product $r_q^2$ but rather the number of combinations with repetition of $r_q$ things taken two at a time.

For $m$ lines at the root, suppose the rooted trees added are specified as to number of points by the $m$-part partition

$$(1^{k_1} 2^{k_2} \cdots (n-1)^{k_{n-1}})$$

so that $k_1 + k_2 + \cdots + k_{n-1} = m$, and $k_1 + 2k_2 + \cdots + (n-1)k_{n-1} = n - 1$; then the number of distinct possibilities corresponding to this partition is

$$\binom{r_1 + k_1 - 1}{k_1}\binom{r_2 + k_2 - 1}{k_2} \cdots \binom{r_{n-1} + k_{n-1} - 1}{k_{n-1}}$$

each factor representing the number of possibilities for a repeated part.

Finally, $r_n$ is the sum of all such numbers over all partitions of $n - 1$; that is,

$$r_n = \Sigma \binom{r_1 + k_1 - 1}{k_1}\binom{r_2 + k_2 - 1}{k_2} \cdots \binom{r_{n-1} + k_{n-1} - 1}{k_{n-1}} \tag{41}$$

with $k_1 + 2k_2 + \cdots + nk_{n-1} = n - 1$. This is simpler in the generating-function form, since

$$\begin{aligned} r(x) &= r_1x + r_2x^2 + \cdots \\ &= x\,\Sigma \binom{r_1 + k_1 - 1}{k_j} x^{k_1} \cdots \Sigma \binom{r_j + k_j - 1}{k_j} x^{jk_j} \cdots \\ &= x(1-x)^{-r_1}(1-x^2)^{-r_2} \cdots \end{aligned} \tag{42}$$

which is Cayley's result. As noted in the introduction to this chapter, it is markedly similar to (1), the generating function for partitions. The generating function $r(x)$ is the enumerator for rooted trees by number of points, all points other than the root being alike.

An alternate form due to Pólya, **20**, is worth noticing for later work. Take the logarithm of (42); then

$$\begin{aligned} \log r(x) &= \log x + \Sigma r_j \log (1 - x^j)^{-1} \\ &= \log x + \Sigma\, r_j(x^j + x^{2j}/2 + x^{3j}/3 + \cdots) \\ &= \log x + r(x) + r(x^2)/2 + r(x^3)/3 + \cdots \end{aligned}$$

or

$$r(x) = x \exp\,[r(x) + r(x^2)/2 + r(x^3)/3 + \cdots] \tag{43}$$

The same procedure may be used for rooted trees with unlike points, that is, for completely point-labeled rooted trees. Take $R_n$ for the number with $n$ points, and $R_n(m)$ for the number with $m$ lines at the root. Note first that the label for the root may be chosen in $n$ ways and, for any particular choice, $n - 1$ labels are available for the remaining points.

Then, following Pólya, **20**,

$$\begin{aligned} R_n(1) &= nR_{n-1} \\ R_n(2) &= \frac{n}{2!}\sum_{j=1}^{n-2}\binom{n-1}{j} R_jR_{n-1-j} \end{aligned}$$

since, in the latter, the labels for the (rooted) trees added to the two lines at the root may be chosen in $\binom{n-1}{j}$ ways when one of the added trees has $j$ points. Note that, because of the labels, the added trees are never alike. The factor 2! is included to remove duplications obtained by permutation of the lines at the root.

With $R_0 = 0$, these results may be written:

$$R_n(1) = nR^{n-1}, \qquad R^j \equiv R_j$$
$$2!R_n(2) = n(R + R)^{n-1}, \qquad R^j \equiv R_j$$

from which it is an easy guess that

$$m!\, R_n(m) = n(R + R + \cdots + R)^{n-1}, \qquad R^j \equiv R_j \tag{44}$$

the expression on the right containing $m$ terms.

Proof of (44) is immediate once it is noted that the number of ways of assigning $n - 1$ labels to trees with $p, q, r, \cdots$ points is the multinomial coefficient $(n - 1)!/p!q!r! \cdots$

If the enumerator of the numbers $R_n$ is taken in the exponential form

$$R(x) = R_1x + R_2x^2/2! + R_3x^3/3! + \cdots$$

then, from $R_1 = 1$ and

$$R_n = \sum_{m=1} R_n(m), \qquad n > 1$$

it follows at once that

$$\begin{aligned} R(x) &= x + xR(x) + xR^2(x)/2! + \cdots \\ &= x \exp R(x) \end{aligned} \tag{45}$$

a result obtained by Pólya, **20**.

From the short enumeration in Fig. 3, it is found that $R_n = n^{n-1}, n \leq 5$, and in fact the solution of (45) which vanishes for $x = 0$ is

$$R(x) = x + 2x^2/2! + 3^2x^3/3! + \cdots + n^{n-1}x^n/n! + \cdots$$

so that

$$R_n = n^{n-1} \tag{46}$$

which is Cayley's formula.

For this case of distinct points, trees and rooted trees are not essentially different. For each (free) tree with $n$ distinct points corresponds to $n$ distinct rooted trees, and, if $T_n$ is the number of trees with $n$ distinct points,

$$T_n = n^{-1}R_n = n^{n-2} \tag{47}$$

a result again known to Cayley.

These results are readily extended to oriented trees. A tree is oriented when each of its lines is assigned a direction for which there are just two choices; for pictorial purposes, this is usually indicated by an arrowhead.

The numbers for the cases of distinct points follow at once from the remarks that (i) $n - 1$ lines (in trees with $n$ points) may be oriented in $2^{n-1}$ ways and (ii) each such orientation of a free or rooted tree with distinct points is distinct. Hence, if $P_n$ and $Q_n$ are respectively the numbers of oriented rooted and free trees with distinct points, then

$$P_n = 2^{n-1}R_n = (2n)^{n-1} \tag{48}$$

$$Q_n = 2^{n-1}T_n = 2(2n)^{n-2} \tag{49}$$

For the case of like points, take $\rho_n$ for the number of oriented rooted trees with $n$ points, and $\rho_n(m)$ for the corresponding number with $m$ lines at the root. Then first $\rho_n(1) = 2\rho_{n-1}$, since the line at the root has two orientations. In the same way, for $m$ lines at the root, a factor of 2 is associated with the number of oriented trees which may be added to a line at the root, the correspondent to (41) being

$$\rho_n = \Sigma \binom{2\rho_1 + k_1 - 1}{k_1} \binom{2\rho_2 + k_2 - 1}{k_2} \cdots \binom{2\rho_{n-1} + k_{n-1} - 1}{k_{n-1}}$$

Hence, if the enumerator is

$$\rho(x) = \rho_1 x + \rho_2 x^2 + \cdots + \rho_n x^n + \cdots \tag{50}$$

it follows just as for $r(x)$ that

$$\rho(x) = x(1 - x)^{-2\rho_1}(1 - x^2)^{-2\rho_2} \cdots (1 - x^k)^{-2\rho_k} \cdots \tag{51}$$

$$= x \exp 2[\rho(x) + \rho(x^2)/2 + \cdots + \rho(x^k)/k + \cdots] \tag{52}$$

## 8. PÓLYA'S THEOREM

The results given above may be unified and generalized by a theorem discovered by Pólya (l.c.) in his classic study of trees and their relatives. As the theorem contains several distinct notions, and the terminology and statement here differ from those of Pólya, a few preliminary remarks are needed.

The theorem concerns the relations of two enumerators, each with the same number of variables which for convenience of statement is here taken as two.

The first of these, $S(x, y)$, is the enumerator of a store (population or set) of objects according to rank, size, or content with respect to two given characteristics; for example, the trees with some points made distinct by labels according to the number of points and number of labels; $S(x, y)$ may be any two-variable generating function, that is, any function of the form [equation (2.4)]

$$S(x, y) = \Sigma S_{ij} f_i(x) g_j(y)$$

with the sets of functions $(f_i(x))$, $(g_j(y))$ linearly independent. The choice of these functions is dictated by the characteristics in question, just as in Chapter 1 the enumeration of combinations was by ordinary generating functions, whereas permutations were enumerated by exponential generating functions.

The second enumerator, $T(x, y)$, is the enumerator with respect to the same characteristics as the store, of the distinct (non-equivalent) ordered selections of $n$ of the store objects, each of the objects being selected independently; $T(x, y)$ then differs from $S(x, y)$ only in numerical coefficients (which may be taken as $T_{ij}$).

This leaves undefined the nature of the distinctness of a selection of $n$ objects and the relation of the two characteristics of this selection to those for individual objects.

As for the first of these, two selections are distinct if they differ with respect to at least one object. But, furthermore, there may be distinctness with respect to order of selection. This is kept general in the following way. Two ordered selections (of the same objects) are taken as equivalent (non-distinct) if there is a permutation of a group $G$ which sends one into the other. The group $G$ is specified by its *cycle index*

$$H_n(t_1, t_2, \cdots, t_n) = \frac{1}{h} \Sigma h_{i_1 i_2 \ldots i_n} t_1^{i_1} t_2^{i_2} \cdots t_n^{i_n}$$

where $i_1 + 2i_2 + \cdots + ni_n = n$, $h$ is the total number of permutations of $G$, and $h_{i_1 i_2 \ldots i_n}$ is the number having $i_1$ cycles of length one, $i_2$ of length two, and so on. Note that $H_n(1, 1, \cdots, 1) = 1$, and that if all orders of selection are distinct $H_n = t_1^n$, whereas, at the other extreme, if all orders are alike,

$$H_n(t_1, t_2, \cdots, t_n) = C_n(t_1, t_2, \cdots, t_n)/n!$$

where $C_n$ is the cycle indicator of Chapter 4.

As to the second question, the composition of characteristics, the rule is as follows. The characteristics must be such that, when all orders of selection are distinct, the enumerator for selections of $n$, $T_n(x, y)$, satisfies $T_n(x, y) = S^n(x, y)$, as is natural for the use of generating functions. This is easily modified to accommodate selections made from more than one store, but the modification is passed over to avoid further complication. For such characteristics, the enumerator $S(x,y)$ is said to satisfy the *product rule*.

The theorem may now be stated as follows:

**Theorem (Pólya).** *If objects are chosen independently from a store having enumerator* $S \equiv S(x_1, x_2, \cdots)$ *which obeys the product rule, and if the order equivalence of choice is specified by the cycle index* $H_n(t_1, t_2, \cdots t_n)$, *then the enumerator of distinct (inequivalent) choices of* $n$ *is*

$$H_n(S_1, S_2, \cdots, S_n)$$

*where* $S_1 = S$ *and* $S_k$ *is the enumerator for choices of* $k$ *objects which remain invariant under cyclic permutations of length* $k$.

The proof of the theorem is as follows. Suppose the cycle index $H$ is of the order $h$, so that the permutations included in $H$ may be written $H_1, H_2, \cdots, H_h$ and suppose the enumerator for choices invariant under permutation $H_i$ is

$$E(H_i) \equiv E(x_1, x_2 \cdots; H_i)$$

Any particular choice $C$ ($n$ particular objects in a particular order) is invariant for some $g$ of these permutations ($g$ is at least 1 since one of the permutations is the identity); in fact, these $g$ permutations are a subgroup of $H$, and $g$ divides $h$. For all the permutations of $H$ there are $h/g$ choices equivalent to $C$ (including $C$ itself).

Now in the sum

$$E(H_1) + E(H_2) + \cdots + E(H_h)$$

any choice equivalent to $C$ appears in just $g$ terms of the sum (since $g$ permutations leave it invariant). Hence altogether they contribute $g(h/g) = h$ to the sum and

$$[E(H_1) + E(H_2) + \cdots + E(H_h)]/h$$

enumerates the inequivalent choices.

Take $H_i$ as a permutation of cycle structure $1^{i_1}2^{i_2} \cdots$. By the product rule, the enumerator $E(H_i)$ may be examined cycle by cycle and, if the enumerator for cycles of length $k$ is $S_k$,

$$E(H_i) = S_1^{i_1} S_2^{i_2} \cdots$$

Summing on $i$ gives the result in the theorem.

Notice that the only choices invariant for cycles of length $k$ are those for which all objects are alike. If the store enumerator is of the form (for two variables, for simplicity)

$$S(x, y) = \Sigma S_{nm} x^n y^m$$

then

$$S_k(x, y) = \Sigma S_{nm} x^{kn} y^{km} = S(x^k, y^k) \tag{53}$$

since choosing all objects alike multiplies the exponents.

For stores containing objects with permutable characteristics for which, as already mentioned, the enumerator is of the form

$$S(x, y) = \Sigma S_{nm} x^n y^m / m!$$

the choices of like objects are limited to those for which $m = 0$, and

$$S_k(x, y) = \Sigma S_{n0} x^{kn} = S(x^k, 0)$$

The existence of these cases is the reason for stating the conclusion of the theorem at a stage earlier than Pólya.

Two simple instances of the theorem, which use material of preceding chapters, may help to clarify its meaning.

First, if permutations are in question so the cycle index is $t_1^n$ as above, and the store is of $m$ unlike objects so its enumerator is $x_1 + x_2 + \cdots + x_m$, then, by the theorem, the enumerator of permutations with repetition $n$ at a time is $(x_1 + x_2 + \cdots + x_m)^n$, as in Chapter 1.

Next, for combinations with repetition, the store is the same but the cycle index is $C_n(t_1, t_2, \cdots, t_n)/n!$, and, by the theorem, the combinations of $m$ distinct things $n$ at a time and with repetition are enumerated by

$$C_n(x_1 + x_2 + \cdots + x_m, x_1^2 + x_2^2 + \cdots + x_m^2, \cdots, \\ x_1^n + x_2^n + \cdots + x_m^n)/n!$$

This may be verified as follows: The enumerator for such combinations, any number at a time, as given in Chapter 1, is in present notation

$$G(t) = (1 + x_1 t + x_1^2 t^2 + \cdots)(1 + x_2 t + x_2^2 t^2 + \cdots) \cdots \\ (1 + x_m t + x_m^2 t^2 + \cdots)$$

$$= \sum_{n=0}^{\infty} G_n(x_1, x_2, \cdots, x_m) t^n$$

and, if the result of the theorem is correct,

$$G_n(x_1, x_2, \cdots, x_m) = C_n(s_1, s_2, \cdots, s_n)/n!$$

where for brevity

$$s_k = x_1^k + x_2^k + \cdots + x_m^k$$

(Note that $s_k$ is the $k$th power sum of the $m$ variables $x_1$ to $x_m$.)

To show this, note first that

$$\begin{aligned} G(t) &= \exp \log (1 - x_1 t)^{-1} (1 - x_2 t)^{-1} \cdots (1 - x_m t)^{-1} \\ &= \exp (\log (1 - x_1 t)^{-1} + \log (1 - x_2 t)^{-1} + \cdots + \log (1 - x_m t)^{-1}) \\ &= \exp \sum_{k=1}^{m} (x_k t + x_k^2 t^2/2 + \cdots + x_k^n t^n/n + \cdots) \\ &= \exp (s_1 t + s_2 t^2/2 + \cdots + s_n t^n/n + \cdots) \end{aligned} \tag{54}$$

On the other hand, by equation (4.3$a$)

$$\Sigma t^n C_n(s_1, s_2, \cdots, s_n)/n! = \exp(s_1 t + s_2 t^2/2 + \cdots)$$

This proves the result.

A verification of more immediate interest is the use of the theorem to find equation (43), the enumerator identity for rooted trees. This is done as before by considering rooted trees with $n$ lines at the root. The store consists of the rooted trees which may be connected to these lines and has enumerator (by number of points) $r(x)$, using earlier notation. Any permutation of the $n$ lines leaves the choice the same, so the cycle index is $C_n(t_1, t_2, \cdots, t_n)/n!$. Since the root contributes one point, the enumerator for rooted trees with $n$ lines is

$$xC_n(r(x), r(x^2), \cdots, r(x^n))/n!$$

and using equation (4.3) again

$$r(x) = x \exp(r(x) + r(x^2)/2 + \cdots + r(x^k)/k + \cdots)$$

which is equation (43).

A similar procedure verifies equation (52). The single difference appears in the store which has enumerator $2\rho(x)$ since the line at the root has two orientations.

Consider now rooted trees with $p$ points, $m$ of which, including the root, are labeled. A labeled point is distinct from all other labeled points and from the unlabeled points, and the labels of dissimilar points may be permuted. It is convenient to think of the labels as numbers, and to use numeral 1 for the single label, numerals 1 and 2 for two labels, and so on, but this is not necessary; the labels for the several cases may be chosen quite independently. Note that this kind of labeling is distinct from choosing each label from a store of $c$ labels (which might be colors), for which some results appear in the problems.

Let $r_{pm}$ be the number of rooted trees with $p$ points, $m$ of which, including the root are labeled, and let their enumerator be

$$\begin{aligned} r(x, y) &= xr_1(y) + x^2 r_2(y) + \cdots + x^p r_p(y) + \cdots \\ &= \sum_1^\infty x^p \sum_0^p r_{pm} y^m/m! \end{aligned}$$

Note that $y$ is a variable of the permutation kind, and that $r(x, 0) = r(x)$, the function defined above.

For labeled rooted trees with $n$ lines at the root, the store has enumerator $r(x, y)$ and, as before, the cycle index is $C_n(t_1, t_2, \cdots, t_n)/n!$ and no two rooted trees may have the same labels; therefore, since the root has label enumerator $x(1 + y)$

$$r(x, y) = x(1 + y) \exp [r(x, y) + r(x^2)/2 + \cdots + r(x^k)/k + \cdots] \quad (55)$$

For $y = 0$, this becomes (43) and, by using both, the symmetrical form appears:

$$r(x, y) \exp r(x) = (1 + y)r(x) \exp r(x, y) \quad (56)$$

The consequences of this relation which are convenient for computation are the results of taking partial derivatives. Using subscripts to denote the latter ($r_x(x, y) = \partial r(x, y)/\partial x$, etc.), they are as follows:

$$a(x)r_x(x, y) = (1 + y)r_y(x, y) = r(x, y)/(1 - r(x, y)) \quad (57)$$

where for brevity

$$a(x) = r(x)/r'(x)(1 - r(x))$$

the prime denoting an ordinary derivative. Note that $a(x)$ is a power series in $x$; indeed,

$$\begin{aligned} a(x) &= a_1x + a_2x^2 + \cdots + a_nx^n + \cdots \\ &= x - x^3 - x^4 - 2x^5 + x^6 - 3x^7 + 4x^8 - x^9 + x^{10} + \cdots \end{aligned}$$

Then from the first half of (57) it follows that

$$\begin{aligned} (1 + y)r_n'(y) = nr_n(y) - (n - 2)r_{n-2}(y) - (n - 3)r_{n-3}(y) - 2(n - 4)r_{n-4}(y) \\ + \cdots + (n - k)r_{n-k}(y)a_{k+1} + \cdots \quad (58) \end{aligned}$$

from which the values in Table 3 have been computed [$m = 0(1)n$; $n = 1(1)10$].

TABLE 3

NUMBERS OF POINT LABELED ROOTED TREES

| Number of Labels | Number of Points | | | | | | | | | |
|---|---|---|---|---|---|---|---|---|---|---|
| | 1 | 2 | 3 | 4 | 5 | 6 | 7 | 8 | 9 | 10 |
| 0 | 1 | 1 | 2 | 4 | 9 | 20 | 48 | 115 | 286 | 719 |
| 1 | 1 | 2 | 5 | 13 | 35 | 95 | 262 | 727 | 2033 | 5714 |
| 2 | | 2 | 9 | 34 | 119 | 401 | 1316 | 4247 | 13532 | 42712 |
| 3 | | | 9 | 64 | 326 | 1433 | 5799 | 22224 | 81987 | 2 93987 |
| 4 | | | | 64 | 625 | 4016 | 21256 | 1 00407 | 4 39646 | 18 23298 |
| 5 | | | | | 625 | 7776 | 60387 | 3 73895 | 20 19348 | 99 41905 |
| 6 | | | | | | 7776 | 1 17649 | 10 71904 | 76 01777 | 462 05469 |
| 7 | | | | | | | 1 17649 | 20 97152 | 219 35132 | 1753 29789 |
| 8 | | | | | | | | 20 97152 | 430 46721 | 5083 88608 |
| 9 | | | | | | | | | 430 46721 | 10000 00000 |
| 10 | | | | | | | | | | 10000 00000 |

The fully labeled rooted trees are enumerated by $r_{nn}$ and, hence, $r_{nn}$ should be equal to $R_n$ of equation (46), and (56) should imply (45). To show this, set $xy = z$ and

$$r(x, z) = R_0(z) + xR_1(z) + \cdots + x^n R_n(z) + \cdots \tag{59}$$

with

$$R_n(z) = \sum_{m=0}^{\infty} r_{n+m,m} z^m/m!$$

Note that $R_0(z)$ should be $R(z)$ with $R(x)$ defined by (45).

Since (56) becomes

$$r(x, z) \exp r(x) = (x + z)(r(x)/x) \exp r(x, z)$$

and $r(0, z) = R_0(z)$, $r(0) = 0$, $(r(x)/x)_{x=0} = 1$, it follows at once that

$$R_0(z) = z \exp R_0(z)$$

which is (45) in present notation.

For completeness note that

$$\begin{aligned} xa(x)r_x(x, z) &= (x^2 + xz - za(x))r_z(x, z) \\ &= (x^2 + xz - za(x))r(x, z)/(x + z)(1 - r(x, z)) \end{aligned} \tag{60}$$

## 9. TREES

To relate enumerators for trees to those for rooted trees, a procedure must be found by which every tree is made to look like (and behave for enumeration like) a rooted tree. This may be done in two ways.

First, every tree has either a center or a bicenter. Stripping off all outer lines of a tree results in a smaller tree with a number of outer lines which may be stripped off in turn; the process continues until it ends in a single point, the center, or a single line, the points of which are together the bicenter. A tree with center looks like a rooted tree with the center as root; a tree with bicenter looks like two rooted trees joined by the bicentral line.

However, the enumerator relationship is easier to find (following Pólya) by the second way which goes as follows. Define the height of any branch of a tree at any point as the number of lines it contains; the outer points all have one branch which contains all lines and, hence, all outer points have height $n - 1$ for a tree with $n$ points. Define the height of a point as the height of the largest of its branches. Then the point with the smallest height is called the centroid. Every tree has either a centroid or a bicentroid; the latter consists of two points of equal height joined by a line (the tree with two points is both a bicentroidal and a bicentral tree).

The two representations of trees, by centers and by centroids, are the same for $n < 5$. However the tree

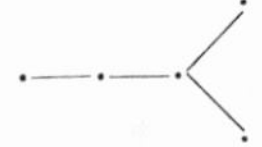

has a bicenter but only a single centroid (the fork point).

If a tree is bicentroidal, it must have two trees with the same number of points (including in each a centroidal point) at the ends of the line joining the two centroids (by definition). Hence, for an odd number of points no tree is bicentroidal, and for an even number of points, say $2m$, the terminating trees may be regarded as rooted trees each with $m$ points.

For an odd number of points, the tree with no forks has a centroid of height $(n - 1)/2$, and no tree with the same number of points can have a centroid of greater height. For a tree with an even number of points, the maximum height of a centroid is $[(n - 1)/2]$, with the brackets as usual indicating the integral part of the number enclosed. A tree with this height may be obtained, for example, by adding a line at the centroid to the tree just mentioned with no forks and one less point, and it can be proved that this is the maximum height.

Let $t'(x, y)$ be the enumerator for labeled centroidal trees, $t''(x, y)$ the enumerator for labeled bicentroidal trees; then

$$t(x, y) = t'(x, y) + t''(x, y)$$

is the enumerator for labeled trees. Let $t_p'(y)$, $t_p''(y)$, and $t_p(y)$ be the corresponding enumerators by labels of trees with $p$ points.

The bicentroidal trees are enumerated at once by the remarks above and by Pólya's theorem. First

$$t''_{2q+1}(y) = 0 \tag{61}$$

and, since the centroids are similar, the cycle index is $(t_1^2 + t_2)/2$ and

$$t''_{2q}(y) = (r_q^2(y) + r_q)/2 \tag{62}$$

The centroidal trees are included in the enumeration of rooted trees (classified by the number of lines at the root) but the latter enumeration includes trees of heights greater than $[(n - 1)/2]$. For example, if a rooted tree with $n$ points and $m$ lines at the root, $m_1$ of which are terminated by rooted trees with one point, $m_2$ with two points, etc., is enumerated by $r_{[m]}(y)$, where $[m]$ is a partition $1^{m_1}2^{m_2} \cdots$ of $n - 1$ into $m$ parts, then

$$t_3'(y) = (1 + y)r_{11}(y)$$

while

$$r_3(y) = (1 + y)r_2(y) + (1 + y)r_{11}(y)$$

Noting that $r_1(y) = 1 + y$, it follows that

$$t_3'(y) = r_3(y) - r_2(y)r_1(y)$$

Again

$$t_4'(y) = (1 + y)r_{111}(y)$$

$$r_4(y) = (1 + y)r_3(y) + (1 + y)r_{21}(y) + (1 + y)r_{111}(y)$$

and

$$t_4'(y) = r_4(y) - r_3(y)r_1(y) - (1 + y)r_{21}(y)$$
$$= r_4(y) - r_3(y)r_1(y) - r_2^2(y)$$

The last step uses $r_{21}(y) = r_2(y)r_1(y)$ and $r_1^2(y) = r_2(y)$, the first of which follows from the fact that the two branches at the root are unlike and the second from $r_2(y) = (1 + y)r_1(y)$.

By similar reductions it turns out that

$$t_{2q}'(y) = r_{2q}(y) - r_{2q-1}(y)r_1(y) - r_{2q-2}(y)r_2(y) - \cdots - r_q^2(y) \tag{63}$$

$$t_{2q+1}'(y) = r_{2q+1}(y) - r_{2q}(y)r_1(y) - r_{2q-1}(y)r_2(y) - \cdots - r_{q+1}(y)r_q(y) \tag{64}$$

Summing $x^p t_p'(y)$ and $x^p t_p''(y)$ on $p$ shows that

$$t(x, y) = r(x, y) - \tfrac{1}{2}\, r^2(x, y) + \tfrac{1}{2}\, r(x^2) \tag{65}$$

The term $r^2(x, y)/2$ comes from the products $r_{p-k}(y)r_k(y)$, and the term $r(x^2)/2$ comes from the part $r_q/2$ of $t_{2q}''(y)$.

Setting $y = 0$ and $t(x, 0) = t(x)$ leads to

$$t(x) = r(x) - \tfrac{1}{2}\, r^2(x) + \tfrac{1}{2}\, r(x^2) \tag{66}$$

a result first obtained by Otter, **19**.

For other verifications, note first that, using suffix notation for partial derivatives as before,

$$a(x)t_x(x, y) = r(x, y) + xa(x)r'(x^2) \tag{67}$$

$$(1 + y)t_y(x, y) = r(x, y) \tag{68}$$
$$= a(x)(t_x(x, y) - xr'(x^2))$$

Here $a(x)$ is the function just after equation (57), $(a(x) = r(x)/r'(x)$ $(1 - r(x))$, and it may be noticed that from (66)

$$t'(x) - xr'(x^2) = r'(x)(1 - r(x))$$
$$= r(x)/a(x) \tag{69}$$

The prime denotes a derivative.

From (68) with $y = 0$, it is found that

$$t_y(x, 0) = \Sigma x^p t_{p1} = r(x)$$

TABLE 4

THE NUMBER OF TREES, ROOTED TREES, ORIENTED TREES, AND ORIENTED ROOTED TREES WITH $n$ POINTS

| $n$ | 1 | 2 | 3 | 4 | 5 | 6 | 7 | 8 | 9 | 10 | 11 | 12 |
|---|---|---|---|---|---|---|---|---|---|---|---|---|
| $t_n$ | 1 | 1 | 1 | 2 | 3 | 6 | 11 | 23 | 47 | 106 | 235 | 551 |
| $r_n$ | 1 | 1 | 2 | 4 | 9 | 20 | 48 | 115 | 286 | 719 | 1842 | 4766 |
| $\tau_n$ | 1 | 1 | 3 | 8 | 27 | 91 | 350 | 1376 | 5743 | 24635 | 1 08968 | 4 92180 |
| $\rho_n$ | 1 | 2 | 7 | 26 | 107 | 458 | 2058 | 9498 | 44947 | 2 16598 | 10 59952 | 52 51806 |

| $n$ | 13 | 14 | 15 | 16 | 17 |
|---|---|---|---|---|---|
| $t_n$ | 1301 | 3159 | 7741 | 19320 | 48629 |
| $r_n$ | 12486 | 32973 | 87811 | 2 35381 | 6 34847 |
| $\tau_n$ | 22 66502 | 105 98452 | 502 35931 | 2408 72654 | 11667 32814 |
| $\rho_n$ | 262 97238 | 1328 56766 | 6763 98395 | 34667 99104 | 1 78738 08798 |

| $n$ | 18 | 19 | 20 | 21 |
|---|---|---|---|---|
| $t_n$ | 1 23867 | 3 17955 | 8 23065 | 21 44505 |
| $r_n$ | 17 21159 | 46 88676 | 128 26228 | 352 21832 |
| $\tau_n$ | 56820 01435 | 4 80687 87314 | 13 93549 22608 | 69 58085 54300 |
| $\rho_n$ | 9 26300 98886 | 48 22926 84506 | 252 16101 75006 | 1323 35730 19372 |

| $n$ | 22 | 23 | 24 | 25 | 26 |
|---|---|---|---|---|---|
| $t_n$ | 56 23756 | 148 28074 | 392 99897 | 1046 36890 | 2797 93450 |
| $r_n$ | 970 55181 | 2682 82855 | 7437 24984 | 20671 74645 | 57596 36510 |

TABLE 5

NUMBER OF POINT LABELED TREES

| Number of Labels | Number of Points | | | | | | | | | |
|---|---|---|---|---|---|---|---|---|---|---|
| | 1 | 2 | 3 | 4 | 5 | 6 | 7 | 8 | 9 | 10 |
| 0 | 1 | 1 | 1 | 2 | 3 | 6 | 11 | 23 | 47 | 106 |
| 1 | 1 | 1 | 2 | 4 | 9 | 20 | 48 | 115 | 286 | 719 |
| 2 | | 1 | 3 | 9 | 26 | 75 | 214 | 612 | 1747 | 4995 |
| 3 | | | 3 | 16 | 67 | 251 | 888 | 3023 | 10038 | 32722 |
| 4 | | | | 16 | 125 | 680 | 3135 | 13155 | 51873 | 1 95821 |
| 5 | | | | | 125 | 1296 | 8716 | 47787 | 2 32154 | 10 40014 |
| 6 | | | | | | 1296 | 16807 | 1 34960 | 8 58578 | 47 41835 |
| 7 | | | | | | | 16807 | 2 62144 | 24 50309 | 177 54459 |
| 8 | | | | | | | | 2 62144 | 47 82969 | 510 48576 |
| 9 | | | | | | | | | 47 82969 | 1000 00000 |
| 10 | | | | | | | | | | 1000 00000 |

which implies that $t_{p1} = r_p$, a verification, since rooted trees by definition are trees with one exceptional (labeled) point, the root.

Also, making the substitution $z = xy$ and writing (as for rooted trees)

$$t(x, z) = T_0(z) + xT_1(z) + \cdots$$

with

$$T_n(z) = \sum_{m=0}^{\infty} t_{n+m,m} z^m/m! \tag{70}$$

it follows from

$$t_z(x, z) = r_z(x, z)(1 - r(x, z))$$

and from equation (60) that

$$(x + z)t_z(x, z) = r(x, z) \tag{71}$$

so that

$$T'_{n-1}(z) + zT'_n(z) = R_n(z) \tag{72}$$

and, in particular,

$$zT'_0(z) = R_0(z)$$

The last implies $nt_{nn} = r_{nn}$, or, in the notation of Section 7, $nT_n = R_n$.

The relation of enumerators $\tau(x, y)$ and $\rho(x, y)$ for oriented trees and oriented rooted trees respectively may be obtained almost exactly as above, the single important difference being that the orientation of the line between the two centroids of a bicentroidal tree removes the symmetry of the centroids and

$$\tau''_{2q}(y) = \rho_q^2(y) \tag{73}$$

The relation turns out to be

$$\tau(x, y) = \rho(x, y) - \rho^2(x, y) \tag{74}$$

Table 4 shows the numbers $r_n$ and $t_n$ for $n = 1(1)26$, $\rho_n$ and $\tau_n$ for $n = 1(1)21$. Table 5 gives $t_{nm}$ for $n = 1(1)10$ and $m = 0(1)n$.

## 10. SERIES-PARALLEL NETWORKS

P. A. MacMahon, **18**, used the ideas and methods introduced by Cayley for trees to enumerate what he called combinations of resistances and what are now called two-terminal series-parallel networks. These are electric networks considered topologically or geometrically, that is, without the electrical properties of the elements connected. The two-terminal networks have great similarity to rooted trees, both formally and in their enumerators, because the terminals like the root are special points and one terminal may be regarded as the merging of all outer points of a rooted tree.

But the networks differ from trees in two ways. First, in a tree adjacent points are connected by just one line, whereas in a network any number of lines are permitted. Next, networks are regarded as equivalent, not only topologically (like trees) but also when interchange of elements in series brings them into congruence; otherwise stated, series interchange is an equivalence operation.

The two-terminal series-parallel networks with four or less elements (lines) are shown in Fig. 4. It will be noticed that, for more than one

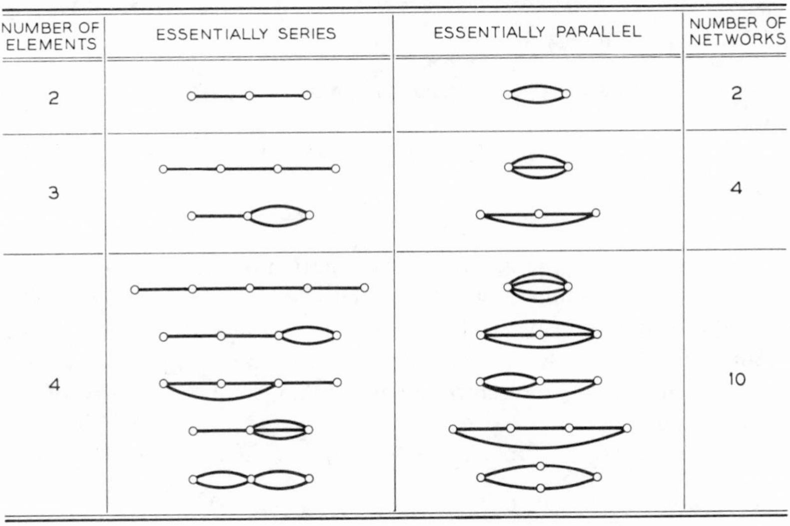

| NUMBER OF ELEMENTS | ESSENTIALLY SERIES | ESSENTIALLY PARALLEL | NUMBER OF NETWORKS |
|---|---|---|---|
| 2 | | | 2 |
| 3 | | | 4 |
| 4 | | | 10 |

**Fig. 4.** Series-parallel two-terminal networks.

element, the number of networks is even. This is a consequence of a duality property. The total may be divided, as in the figure, into two classes, called *essentially series* and *essentially parallel*; any network in one class has a mate or dual in the other which is obtained simply by interchanging the words "series" and "parallel" in its verbal description. Essentially series networks are formed by connecting essentially parallel networks in series, and vice versa.

The enumeration of these networks makes use of duality as follows. Take $a_{n,m}$ as the number of essentially parallel networks with $n$ elements, $m$ of which are labeled with distinct labels, and

$$a(x, y) = xa_1(y) + x^2a_2(y) + \cdots \tag{75}$$

with

$$a_n(y) = a_{n0} + a_{n1}y + a_{n2}y^2/2 + \cdots + a_{nn}y^n/n!$$

as the enumerator. Let $S_{nm}$ and $S(x, y)$ have the same meanings for (two-terminal) series-parallel networks.

Then $S_1(y) = a_1(y) = 1 + y$, and, by duality, $S_n(y) = 2a_n(y)$ for $n$ greater than 1, so

$$S(x, y) = 2a(x, y) - x(1 + y) \tag{76}$$

Now consider the networks obtained by connecting $k$ essentially series networks in parallel in the light of Pólya's theorem. The store consists of all essentially series networks and by duality has enumerator $a(x, y)$. Position in parallel is immaterial, so the cycle index is $C_k(t_1, t_2, \cdots, t_k)/k!$. Since no two labeled networks can be alike, $t_j$ is replaced for the enumeration by $a(x^j, 0) \equiv a(x^j)$.

Summing on $k$ gives at once

$$1 + S(x, y) = \exp\,(a(x, y) + a(x^2)/2 + \cdots + a(x^k)/k + \cdots) \tag{77}$$

With $y = 0$, equation (77) corresponds to the enumerator identity found by MacMahon; that is, with $S(x, 0) = S(x)$,

$$1 + S(x) = \exp\,(a(x) + a(x^2)/2 + \cdots + a(x^k)/k + \cdots) \tag{77a}$$

is equivalent to MacMahon's form

$$1 + S(x) = (1 - x)^{-a_1}(1 - x^2)^{-a_2} \cdots (1 - x^k)^{-a_k} \cdots$$

with, of course,

$$a(x) = xa_1 + x^2a_2 + \cdots$$

Using (77a), (77) may be written in the symmetrical form

$$(1 + S(x, y)) \exp a(x) = (1 + S(x)) \exp a(x, y)$$

or finally, by use of (76),

$$(1 + S(x, y)) \exp S(x)/2 = (1 + S(x)) \exp\,(S(x, y) + xy)/2 \tag{78}$$

It follows, just as for rooted trees and with the same notation for partial derivatives, that

$$S_x(x, y) = (y + d(x))(1 + S)/(1 - S) \tag{79}$$

$$S_y(x, y) = x(1 + S)/(1 - S) \tag{80}$$

and

$$xS_x(x, y) = (y + d(x))S_y(x, y) \tag{81}$$

with

$$\begin{aligned} d(x) &= S'(x)(1 - S(x))/(1 + S(x)) \\ &= d_0 + d_1x + d_2x^2 + \cdots + d_nx^n + \cdots \\ &= 1 + 2x + 2x^2 + 6x^3 + 2x^4 + 18x^5 + 2x^6 + 46x^7 + 14x^8 + \cdots \end{aligned}$$

Equation (81), with known $S(x)$ and, hence, $d(x)$, gives the simplest computing recurrence, namely,

$$nS_n(y) = (d_0 + y)S_n'(y) + \sum_1^{n-1} d_k S_{n-k}'(y) \tag{81a}$$

Table 6 shows $S_{nm}$ for $m = 0(1)n$ and $n = 1(1)10$.

TABLE 6

SERIES-PARALLEL NUMBERS $S_{nm}$

Number of Elements, $n$

| $m$ | 1 | 2 | 3 | 4 | 5 | 6 | 7 | 8 | 9 | 10 |
|---|---|---|---|---|---|---|---|---|---|---|
| 0 | 1 | 2 | 4 | 10 | 24 | 66 | 180 | 522 | 1532 | 4624 |
| 1 | 1 | 2 | 6 | 18 | 58 | 186 | 614 | 2034 | 6818 | 22970 |
| 2 | | 2 | 8 | 34 | 136 | 538 | 2080 | 7970 | 30224 | 1 13874 |
| 3 | | | 8 | 52 | 288 | 1424 | 6648 | 29700 | 1 28800 | 5 45600 |
| 4 | | | | 52 | 472 | 3224 | 18888 | 1 01340 | 5 11120 | 24 65904 |
| 5 | | | | | 472 | 5504 | 44712 | 3 02096 | 18 28016 | 102 47424 |
| 6 | | | | | | 5504 | 78416 | 7 38448 | 56 45312 | 379 88096 |
| 7 | | | | | | | 78416 | 13 20064 | 141 38976 | 1205 63808 |
| 8 | | | | | | | | 13 20064 | 256 37824 | 3077 75648 |
| 9 | | | | | | | | | 256 37824 | 5642 75648 |
| 10 | | | | | | | | | | 5642 75648 |

Also, with $xy = z$ and

$$S(x, z) = A_0(z) + xA_1(z) + \cdots + x^n A_n(z) + \cdots \tag{82}$$
$$A_n(z) = \Sigma S_{n+m,m} z^m/m!$$

it follows from (78) that

$$(1 + S(x, z)) \exp S(x)/2 = (1 + S(x)) \exp (S(x, z) + z)/2 \tag{83}$$

$$S_x(x, z) = d(x)(1 + S(x, z))/(1 - S(x, z)) \tag{84}$$
$$= d(x)S_z(x, z)$$

From (83) with $x = 0$ $[S(0) = 0, S(0, z) = A_0(z)]$,

$$1 + A_0(z) = \exp (A_0(z) + z)/2 \tag{85}$$

or, with $A_n = S_{nn}$, by equation (2.39) and the definition of Bell polynomials following equation (2.41),

$$A_n = Y_n(\alpha_1, \alpha_2, \cdots, \alpha_n) \tag{86}$$

with $\alpha_1 = (1 + A_1)/2$, $\alpha_n = A_n/2$, $n > 1$ (the numbers $\alpha_n \equiv a_{nn}$ are the numbers of fully labeled essentially series, or essentially parallel, networks). Equation (86) is equivalent to a relation given by W. Knödel, **16**.

However the numbers $A_n$ are more easily obtained from (85), as follows. First drop the subscript for ease of writing and differentiate; then with a prime denoting a derivative

$$A'(z) = \frac{1 + A(z)}{1 - A(z)}$$

Differentiating again leads to

$$A''(z) = 2\,\frac{1 + A(z)}{[1 - A(z)]^3}$$

and, if it is assumed that the $n$th derivative is of the form

$$A^{(n)}(z) = \frac{1 + A(z)}{[1 - A(z)]^{2n-1}} P_n[A(z)]$$

then the results above and differentiation show that $P_1(x) = 1$, $P_2(x) = 2$, and

$$P_{n+1}(x) = [2n + (2n - 2)x]P_n(x) + (1 - x^2)P'_n(x) \tag{87}$$

Since $A(0) = 0$, it follows that

$$A_n = A^{(n)}(0) = P_n(0)$$

The first few values of the polynomial $P_n(x)$ are

$$\begin{aligned} P_1 &= 1 \qquad & P_3 &= 8 + 4x \\ P_2 &= 2 \qquad & P_4 &= 52 + 56x + 12x^2 \end{aligned}$$

Finally, it may be noticed that (84) is equivalent to the recurrence relation

$$(n + 1)A_{n+1}(z) = \sum_0^n d_k A'_{n-k}(z) \tag{88}$$

from which, in turn, is found

$$(n + 1)S_{n+1+m,m} = \sum_0^n d_k S_{n-k+m+1,m+1} \tag{89}$$

which enables all numbers to be expressed in terms of $d_k$ and $S_{nn} = A_n$.

## 11. LINEAR GRAPHS

A linear graph for present purposes consists of a set of $n$ points and the lines connecting them in pairs so that each pair is either not connected or connected by just one line. The set of $n$ separated points is a linear graph with no lines. Notice that "slings", that is, loops connecting points to themselves, and lines in parallel between points are not permitted. This is for convenience of enumeration; as mentioned above, both are included in the general theory of graphs, as for example in König, **15**.

The enumeration is, of course, by Pólya's theorem, which was invented for the purpose, and is of topologically distinct linear graphs with $n$ points by number of lines.

The store of objects is that of lines to be connected between points and, hence, has enumerator $1 + x$.

The topological distinctness determines the cycle index, which is that of all permutations of the $N = \binom{n}{2}$ pairs of points induced by all permutations of the points themselves. If $g_k$ is the variable indicating a cycle of pairs of points of length $k$, then the cycle index is some function $G_n(g_1, g_2, \cdots, g_n)$. This may be determined from the cycle index for all permutations of the points, namely $C_n(t_1, t_2, \cdots, t_n)/n!$, as will now be shown.*

First, the identity permutation $t_1^n$ corresponds to $g_1^N$, since all pairs of points are unchanged. Next, the cycle of length $n$, $t_n$, corresponds to $g_n^{N/n}$ for $n$ odd and to $g_m g_{2m}^{m-1}$ for $n = 2m$ (note that $m + 2m(m - 1) = N$). For example, with $n = 3$, the cycle (123) of points corresponds to the cycle (12, 23, 13) of pairs of points, which is also of length 3, whereas for $n = 4$, the cycle (1234) corresponds to (12, 23, 34, 14) (13, 24), indicated by $g_4 g_2$.

To show this result for any $n$, note first that a cycle of $n$ points displaces every point, hence corresponds to a permutation of point pairs, each of which is also displaced. The point cycle may be taken at pleasure; suppose it is $(123 \cdots n)$. Then for $k$ any of the numbers 2 to $n$, there is a cycle of pairs

$$(1k; 2, k + 1; \cdots; n - k + 1, n; 1, n - k + 2; \cdots; k - 1, n)$$

This is a cycle of length $n$ unless the pairs $1, k$ and $1, n - k + 2$ are the same, when the length is $n - k + 1$. But if $k = n - k + 2$, $2k = n + 2$ and $n$ must be even; if $n = 2m$, then $k = m + 1$, and the cycle length is $2m - (m + 1) + 1 = m$. Hence, summarizing, for $n$ odd, the pair cycle length is $n$ for every $k$, and the indicator is $g_n^{N/n}$ as stated, since every pair appears in some cycle of $n$. For $n = 2m$, on the other hand, there is just one cycle of length $m$ (for $k = m + 1$) and there must be $m - 1$ cycles of length $n$ to exhaust all pairs.

For a point cycle of length less than $n$, the same results are obtained for the cycles of pairs of points both of which are in this cycle.

Thus, the only case left is that of pair cycles in which one point of the

* The derivation follows unpublished work by my colleague, D. Slepian.

pair belongs to a cycle of length $i$, the other to a cycle of length $j$. A little consideration shows that the index for such cycles is

$$g_{[i,j]}^{(i,j)}$$

with $[i,j]$ the least common multiple and $(i,j)$ the greatest common divisor of $i$ and $j$. Note that $[i,j](i,j) = ij$, the total number of pairs of points, one from each cycle. When $i$ and $j$ are alike, this becomes $g_i^i$.

If $t_i \circ t_j$ indicates the operation just described, the cycle index for pairs is completely determined by the correspondence rules

$$t_{2m} \sim g_m g_{2m}^{m-1}$$

$$t_{2m+1} \sim g_{2m+1}^{m}$$

$$t_i \circ t_j \sim g_M^D, \qquad M = [i,j], \qquad D = (i,j)$$

It must be remembered that the last rule must be applied to all possible point cycle pairs. For example, $t_2 t_1^2$ is transformed by

$$(t_2)(t_1)(t_1)(t_2 \circ t_1)(t_2 \circ t_1)(t_1 \circ t_1)$$

into

$$g_1(1)(1)(g_2)(g_2)(g_1) = g_1^2 g_2^2$$

Also

$$t_{2m}^k \sim (t_{2m})^k (t_{2m} \circ t_{2m})^{\binom{k}{2}} \sim (g_m g_{2m}^{m-1})^k g_{2m}^{2m\binom{k}{2}}$$

$$t_{2m+1}^k \sim (t_{2m+1})^k (t_{2m+1} \circ t_{2m+1})^{\binom{k}{2}} \sim g_{2m+1}^M, \qquad M = \frac{1}{2m+1}\binom{2mk+k}{2}$$

which are convenient working rules.

The first few values of $G_n$ are as follows:

$$G_2 = g_1$$

$$6G_3 = g_1^3 + 3g_1g_2 + 2g_3$$

$$24G_4 = g_1^6 + 9g_1^2g_2^2 + 8g_3^2 + 6g_2g_4$$

$$120G_5 = g_1^{10} + 10g_1^4g_2^3 + 15g_1^2g_2^4 + 20g_1g_3^3 + 20g_1g_3g_6 + 30g_2g_4^2 + 24g_5^2$$

By Pólya's theorem, the enumerator of distinct linear graphs with $n$ points by number of lines, $L_n(x)$, is

$$L_n(x) = G_n(1 + x, 1 + x^2, \cdots, 1 + x^n) \tag{90}$$

Thus

$$L_2(x) = 1 + x$$

$$L_3(x) = [(1 + x)^3 + 3(1 + x)(1 + x^2) + 2(1 + x^3)]/6 = 1 + x + x^2 + x^3$$

$$L_4(x) = [(1 + x)^6 + 9(1 + x)^2(1 + x^2)^2 + 8(1 + x^3)^2 + 6(1 + x^2)(1 + x^4)]/24$$

$$= 1 + x + 2x^2 + 3x^3 + 2x^4 + x^5 + x^6$$

in agreement with the pictorial enumeration in Fig. 1.

Notice that for each $n$

$$L_n(x) = x^N L_n(x^{-1}), \qquad N = \binom{n}{2}$$

This is a consequence of the duality property of the graphs obtained by interchanging the words connected and disconnected in the verbal description of a graph in terms of the condition of its point pairs.

It should be noted that actual computation of the enumerators $L_n(x)$ from equation (90) increases in difficulty almost exponentially with $n$, although it is not difficult to make a table of $G_n$ from that of $C_n$ in Chapter 4. Table 7 shows all values of $L_{nk}$, the distinct linear graphs with $n$ points and $k$ lines, for $n = 2(1)9$, abridging some values for 7, 8, and 9 by the duality property. The values for 8 and 9 have been obtained by indirect methods too involved for statement here.

TABLE 7

THE NUMBER OF LINEAR GRAPHS WITH $n$ POINTS AND $k$ LINES

| $k \backslash n$ | 2 | 3 | 4 | 5 | 6 | 7 | 8 | 9 |
|---|---|---|---|---|---|---|---|---|
| 0 | 1 | 1 | 1 | 1 | 1 | 1 | 1 | 1 |
| 1 | 1 | 1 | 1 | 1 | 1 | 1 | 1 | 1 |
| 2 | | 1 | 2 | 2 | 2 | 2 | 2 | 2 |
| 3 | | 1 | 3 | 4 | 5 | 5 | 5 | 5 |
| 4 | | | 2 | 6 | 9 | 10 | 11 | 11 |
| 5 | | | 1 | 6 | 15 | 21 | 24 | 25 |
| 6 | | | 1 | 6 | 21 | 41 | 56 | 63 |
| 7 | | | | 4 | 24 | 65 | 115 | 148 |
| 8 | | | | 2 | 24 | 97 | 221 | 345 |
| 9 | | | | 1 | 21 | 131 | 402 | 771 |
| 10 | | | | 1 | 15 | 148 | 663 | 1637 |
| 11 | | | | | 9 | 148 | 980 | 3252 |
| 12 | | | | | 5 | 131 | 1312 | 5995 |
| 13 | | | | | 2 | 97 | 1557 | 10120 |
| 14 | | | | | 1 | 65 | 1646 | 15615 |
| 15 | | | | | 1 | 41 | 1557 | 21933 |
| 16 | | | | | | 21 | 1312 | 27987 |
| 17 | | | | | | 10 | 980 | 32403 |
| 18 | | | | | | 5 | 663 | 34040 |

Another use of the theorem is in relating linear graphs and their connected parts. To see this, write first

$$L(x, y) = \sum_{p=1} y^p L_p(x) = \Sigma L_{pk} x^k y^p$$

and then for the distinct labels of points as considered for the trees above, write

$$L(x, y, z) = \Sigma L_{pkj} x^k y^p z^j / j!$$

for the enumerator of linear graphs by number of points, lines, and (point) labels. Write $C(x, y, z)$ as the enumerator for connected graphs, and consider the graphs with $n$ connected parts.

The store is that of connected graphs and has enumerator $C(x, y, z,)$ which satisfies the product rule. The cycle index is $C_n(t_1, t_2, \cdots, t_n)/n!$. Hence, just as for series-parallel networks,

$$1 + L(x, y, z) = \exp [C(x, y, z) + C(x^2, y^2)/2 + \cdots + C(x^k, y^k)/k + \cdots] \tag{91}$$

where for brevity $C(x^k, y^k, 0) \equiv C(x^k, y^k)$.

For $z = 0$, this is a relation found by Harary, **12**, for unlabeled graphs; combining the two leads to $(L(x, y, 0) \equiv L(x, y))$

$$[1 + L(x, y, z)] \exp C(x, y) = [1 + L(x, y)] \exp C(x, y, z) \tag{92}$$

a close relative of (78), and it may be noted that

$$L_z(x, y, z) = (1 + L(x, y, z))C_z(x, y, z) \tag{93}$$

The substitution $yz = w$ leaves (92) and (93) formally unchanged but changes the enumerators into

$$L(x, y, w) = L_0(x, w) + yL_1(x, w) + \cdots$$
$$C(x, y, w) = C_0(x, w) + yC_1(x, w) + \cdots$$

with $L_0(x, w)$ and $C_0(x, w)$, for example, the enumerators for all points labeled. Then from (92) with $y = 0$ and $C(x, 0) = L(x, 0) = 0$, it follows that

$$1 + L_0(x, w) = \exp C_0(x, w) \tag{94}$$

a relation appearing (among others) in a paper by E. N. Gilbert, **9**. Note that for all points labeled the cycle index is $g_1^N$ and

$$L_0(x, w) = \sum_{n=1}^{\infty} (1 + x)^N w^n/n! \qquad N = \binom{n}{2}$$

## 12. CONNECTED GRAPHS WITH ONE CYCLE

The relationships derived just above show the importance of connected graphs in the general theory. The simplest connected graphs are (free) trees. The next simplest are those with a single closed cycle, that is, those graphs consisting of a single polygon with one or more trees spreading from its vertices. These will now be examined since, in addition to pictorial simplicity, they introduce a new kind of cycle index, that of the dihedral group.

They are enumerated as usual by Polya's theorem. Suppose the polygon has $p$ points (and $p$ lines). The store is that of rooted trees which may be connected at these points; hence it has enumerator

$$r(x) = xr_1 + x^2r_2 + \cdots$$

The cycle index is determined by the symmetries of the polygon, as follows.

First for $p = 3$, the triangle, all three vertices are alike; hence, if $D_p(t_1, t_2, \cdots, t_p)$ is the index for $p$ points,

$$D_3(t_1, t_2, t_3) = (t_1^3 + 3t_1t_2 + 2t_3)/6 \tag{95}$$

Next, for $p = 4$, the square, all symmetries are generated by two: (i) the rotation (1234), the numbers referring to the points of the square, which geometrically is a rotation of one right angle, and (ii) the permutation (24), which geometrically is a reflection about a diagonal. The determination of all permutations of its points leaving the square unaltered from these two symmetries is a simple exercise in group theory. Call the two symmetries $R$ and $T$ and take $RT$ as the permutation resulting from $R$ followed by $T$. Notice first that

$$TT = T^2 = I$$
$$RRRR = R^4 = I$$

$I$ standing for the identity permutation $t_1^4$; in words, these mean (i) two reflections are equivalent to none, (ii) four rotations are equivalent to none. The possible simple products of $R$ and $T$ are $RT$, $R^2T$, $R^3T$ and $TR$, $TR^2$ and $TR^3$, but, by simple calculations,

$$RT = TR^3 = (14)(23)$$
$$R^2T = TR^2 = (13)$$
$$R^3T = TR = (12)(34)$$

and any further products may be reduced to one of these or to $T$ or $R^k$, $k = 1, 2, 3$. Hence the complete set of permutations generated by $R$ and $T$ may be taken as

$$I, R, R^2, R^3, T, TR, TR^2, TR^3$$

As mentioned before this is the dihedral group, in this instance of order 8. Noting that $R^2 = (13)(24)$, $R^3 = (1432)$, the cycle index may be written at once as

$$D_4(t_1, t_2, t_3, t_4) = (t_1^4 + 2t_1^2t_2 + 3t_2^2 + 2t_4)/8 \tag{96}$$

For any $p$, the symmetries are generated by two permutations like those above: a rotation $R = (123 \cdots p)$, and a reflection

$$T = \begin{pmatrix} 1\,2\,3 & \cdots p \\ 1\,p\,p-1 & \cdots 2 \end{pmatrix} = (1)(2p)(3, p-1) \cdots (k, p-k+2) \cdots$$

For $p = 2m$, the sequence of transpositions terminates in the unit cycle $m + 1$, and $T$ has cycle structure $1^2 2^{m-1}$; for $p = 2m + 1$, the last transposition is $(m + 1, m + 2)$, and the cycle structure is $1 \cdot 2^m$. For all cases

$$T^2 = R^p = I$$

By direct calculation

$$TRT = (1\,p\,p - 1 \cdots 2) = R^{p-1}$$

or, premultiplying by $T$

$$RT = TR^{p-1}$$

In the language of group theory, $R$ and $R^{p-1}$ are inverses of each other since $R(R^{p-1}) = R^p = I$. They are also transforms of each other since, if $P$ is any permutation and $P\,P_{-1} = I$, $P_{-1}QP = S$, then $S$ is the transform of $Q$. It is important to note that transforms have the same cycle structure.

Iteration of $RT = TR^{p-1}$ shows that $R^kT = TR^{p-k}$ for any $k$, so that the group is of order $2p$ and can be taken as

$$I, R, R^2, \cdots, R^{p-1}, T, TR, TR^2, \cdots, TR^{p-1}$$

The cycle structure may be determined in two parts. First the set $(I, R, \cdots, R^{p-1})$ is a cyclic group of order $p$, the cycle structure of which is given by

$$\Sigma\phi(k)t_k^{p/k}$$

with the sum over all divisors of $p$ and $\phi(k)$ Euler's totient function, the number of numbers less than $k$ and prime to $k$ (this has already appeared in Problem 3.3). Next, consider the set $(T, TR, \cdots, TR^{p-1})$; since

$$R^{p-k}(TR^s)R^k = R^{p-k}TR^{s+k} = TR^kR^{s+k} = TR^{s+2k}$$

by the result noticed above, $TR^s$ and $TR^{s+2k}$ are transforms of each other and have the same cycle structure. For $p$ odd, it follows that all $TR^s$, $s = 0$ to $p - 1$, have the same cycle structure which is that of $T$; hence, the cycle structure of the set is $p\, t_1\, t_2^{(p-1)/2}$. For $p$ even, $q = p/2$ elements $TR^{2k}$ have the structure of $T$ which is $t_1^2\, t_2^{q-1}$, while the remaining $q$ elements $TR^{2k+1}$ have the structure of $TR$ which is $t_2^q$.

TABLE 8

COEFFICIENTS OF POWERS OF ROOTED TREE ENUMERATOR

$$r^m(x) = \Sigma r_n(m) x^{n+m-1}$$

$r_n(m)$

| $m \backslash n$ | 1 | 2 | 3 | 4 | 5 | 6 | 7 | 8 | 9 | 10 |
|---|---|---|---|---|---|---|---|---|---|---|
| 1 | 1 | 1 | 2 | 4 | 9 | 20 | 48 | 115 | 286 | 719 |
| 2 | 1 | 2 | 5 | 12 | 30 | 74 | 188 | 478 | 1235 | 3214 |
| 3 | 1 | 3 | 9 | 25 | 69 | 186 | 503 | 1353 | 3651 | 9865 |
| 4 | 1 | 4 | 14 | 44 | 133 | 388 | 1116 | 3168 | 8938 | 25100 |
| 5 | 1 | 5 | 20 | 70 | 230 | 721 | 2200 | 6575 | 19385 | 56575 |
| 6 | 1 | 6 | 27 | 104 | 369 | 1236 | 3989 | 12522 | 38535 | 1 16808 |
| 7 | 1 | 7 | 35 | 147 | 560 | 1995 | 6790 | 22338 | 71652 | 2 25379 |
| 8 | 1 | 8 | 44 | 200 | 814 | 3072 | 10996 | 37832 | 1 26301 | 4 11824 |
| 9 | 1 | 9 | 54 | 264 | 1143 | 4554 | 17100 | 61407 | 2 13057 | 7 19368 |
| 10 | 1 | 10 | 65 | 340 | 1560 | 6542 | 25710 | 96190 | 3 46360 | 12 09660 |

TABLE 9

NUMBERS $D_{pn}$ OF CONNECTED GRAPHS WITH ONE CYCLE OF LENGTH $p$ AND $n$ POINTS

$D_{pn}$

| $p \backslash n$ | 3 | 4 | 5 | 6 | 7 | 8 | 9 | 10 |
|---|---|---|---|---|---|---|---|---|
| 3 | 1 | 1 | 3 | 7 | 18 | 44 | 117 | 299 |
| 4 | | 1 | 1 | 4 | 9 | 28 | 71 | 202 |
| 5 | | | 1 | 1 | 4 | 10 | 32 | 89 |
| 6 | | | | 1 | 1 | 5 | 13 | 45 |
| 7 | | | | | 1 | 1 | 5 | 14 |
| 8 | | | | | | 1 | 1 | 6 |
| 9 | | | | | | | 1 | 1 |
| 10 | | | | | | | | 1 |

Hence, finally, the cycle index of the dihedral group is given by

$$2pD_p(t_1, t_2, \cdots, t_p) = \Sigma\phi(k) t_k^{p/k} + p\, t_1\, t_2^{(p-1)/2} \tag{97}$$

for $p$ odd, and by

$$2pD_p(t_1, t_2, \cdots, t_p) = \Sigma\phi(k) t_k^{p/k} + q t_2^{q-1}(t_1^2 + t_2) \tag{98}$$

for $p = 2q$. The particular cases (95) and (96) are in agreement.

Now the theorem gives at once the enumerator of connected graphs with one cycle of length $p$ by number of points, $D_p(x)$, as

$$D_p(x) = \Sigma D_{pn} x^n = D_p(r(x), r(x^2), \cdots, r(x^p))/2p \tag{99}$$

For example,

$$D_3(x) = [r^3(x) + 3r^2(x)r(x^2) + 2r(x^3)]/6$$

Table 8 gives the coefficients of $r^m(x)$ for $m = 1(1)10$ which are useful in evaluating such expressions. Table 9 gives the numbers of these graphs.

## REFERENCES


1. L. Carlitz and J. Riordan, The number of labeled two-terminal series—parallel networks, *Duke Math. Journal*, vol. 23 (1956), pp. 435–446.
2. A. Cayley, *Collected Mathematical Papers*, Cambridge, 1889–1897; vol. 3, pp. 242–246; vol. 9, pp. 202–204, 427–460; vol. 11, pp. 365–367; vol. 13, pp. 26–28.
3. E. T. Bell, Interpolated denumerants and Lambert series, *Amer. Journal of Math.*, vol. 65 (1943), pp. 382–386.
4. L. E. Dickson, *History of the Theory of Numbers*, vol. II, Chapter II, Washington, 1920; reprinted New York, 1952.
5. G. W. Ford and G. E. Uhlenbeck, Combinatorial problems in the theory of graphs I, *Proc. Nat. Acad. Sci. USA*, vol. 42 (1956), pp. 122–128.
6. G. W. Ford, R. Z. Norman, and G. E. Uhlenbeck, Combinatorial problems in the theory of graphs II, *Proc. Nat. Acad. Sci. USA*, vol. 42 (1956), pp. 203–208.
7. R. M. Foster, Geometrical circuits of electrical networks, *Trans. Amer. Inst. of Electr. Eng.*, vol. 51 (1932), pp. 309–317.
8. R. M. Foster, The number of series-parallel networks, *Proc. Internat. Congr. Mathematicians*, Cambridge, 1950, vol. 1, p. 646.
9. E. N. Gilbert, Enumeration of labeled graphs, *Canadian Journal of Math.*, vol. 8 (1956), pp. 405–411.
10. H. Gupta, *Tables of Partitions*, Madras, 1939.
11. F. Harary and G. E. Uhlenbeck, On the number of Husimi trees, *Proc. Nat. Acad. Sci. USA*, vol. 39 (1953), pp. 315–322.
12. F. Harary, The number of linear, directed, rooted and connected graphs, *Trans. Amer. Math. Soc.*, vol. 78 (1955), pp. 445–463.
13. ———, Note on the Pólya and Otter formulas for enumerating trees, *Michigan Math. Journal*, vol. 3 (1955–56), pp. 109–112.
14. ———, The number of oriented graphs, *Michigan Math. Journal*, vol. 3 (1957) (to appear).
15. D. König, *Theorie der endlichen und unendlichen Graphen*, Leipzig, 1936; reprinted New York, 1950.
16. W. Knödel, Über Zerfällungen, *Monatsh. Math.*, vol. 55 (1951), pp. 20–27.
17. P. A. MacMahon, *Combinatory Analysis*, vol. II, London, 1916.
18. P. A. MacMahon, The combination of resistances, *The Electrician*, vol. 28 (1892), pp. 601–602.
19. R. Otter, The number of trees, *Annals of Math.*, vol. 49 (1948), pp. 583–599.
20. G. Pólya, Kombinatorische Anzahlbestimmungen für Gruppen, Graphen, und chemische Verbindungen, *Acta Math.*, vol. 68 (1937), pp. 145–253.
21. J. Riordan, The numbers of labeled colored and chromatic trees, *Acta Math.*, vol. 97 (1957), pp. 211–225.
22. J. Riordan and C. E. Shannon, The number of two-terminal series-parallel networks, *Journal of Math. and Physics*, vol. 21 (1942), pp. 83–93.
23. E. Schöder, Vier combinatorische Probleme, *Zeitschrift für Mathematik und Physik*, vol. 15 (1870), pp. 361–376.

## PROBLEMS

1. Show that the denumerant $D(n; da_1, da_2, \cdots, da_m)$, with $d$ the greatest common divisor of $da_1, da_2, \cdots, da_m$, satisfies the following relations

$$D(dn; da_1, da_2, \cdots, da_m) = D(n; a_1, a_2, \cdots, a_m)$$
$$D(dn + e; da_1, da_2, \cdots, da_m) = 0, \qquad e = 1, 2, \cdots, d - 1$$

2. Show that

$$D(5n + m; 1, 5, 10, 25, 50) = D(n; 1, 1, 2, 5, 10), \qquad m = 0, 1, 2, 3, 4$$

Verify the following table:

| $n$ | 0 | 1 | 2 | 3 | 4 | 5 | 6 | 7 | 8 | 9 | 10 | 11 | 12 |
|---|---|---|---|---|---|---|---|---|---|---|---|---|---|
| $D(n; 1, 2)$ | 1 | 1 | 2 | 2 | 3 | 3 | 4 | 4 | 5 | 5 | 6 | 6 | 7 |
| $D(n; 1, 2, 5)$ | 1 | 1 | 2 | 2 | 3 | 4 | 5 | 6 | 7 | 8 | 10 | 11 | 13 |
| $D(n; 1, 2, 5, 10)$ | 1 | 1 | 2 | 2 | 3 | 4 | 5 | 6 | 7 | 8 | 11 | 12 | 15 |
| $D(n; 1, 1, 2, 5, 10)$ | 1 | 2 | 4 | 6 | 9 | 13 | 18 | 24 | 31 | 39 | 50 | 62 | 77 |

| $n$ | 13 | 14 | 15 | 16 | 17 | 19 | 20 |
|---|---|---|---|---|---|---|---|
| $D(n; 1, 2)$ | 7 | 8 | 8 | 9 | 9 | 10 | 11 |
| $D(n; 1, 2, 5)$ | 14 | 16 | 18 | 20 | 22 | 26 | 29 |
| $D(n; 1, 2, 5, 10)$ | 16 | 19 | 22 | 25 | 28 | 34 | 40 |
| $D(n; 1, 1, 2, 5, 10)$ | 93 | 112 | 134 | 159 | 187 | 252 | 292 |

Show that the number of ways of changing a dollar (into pennies, nickels, dimes, quarters and half-dollars) is 292.

3. (*a*) From the identity

$$(1 - x)^{-1} = (1 + x)(1 + x^2)(1 + x^4) \cdots (1 + x^{2^n}) \cdots$$

show that every number has a unique expression in the binary system (base 2).

(*b*) From the same identity, show that every number greater than 1 has as many partitions with an even number of the parts $1, 2, 4, \cdots, 2^n, \cdots$ as with an odd number.

4. (*a*) If

$$H(t, n) = 1/(1 - u)(1 - ut) \cdots (1 - ut^k) \cdots$$
$$= F(t, u)/(1 - u)$$

where $F(t, u)$ is given by equation (6), show that

$$H(t, u) = \Sigma P(t, k)u^k$$

with

$$P(t, k) = p(t, 0) + p(t, 1) + \cdots + p(t, k)$$

and $p(t, k)$ as defined by (7).

(*b*) Show that

$$(1 - u)H(t, u) = H(t, ut)$$
$$(1 - t^k)P(t, k) = P(t, k - 1)$$
$$(1 - t)(1 - t^2) \cdots (1 - t^k)P(t, k) = 1$$

in agreement with (12).

5. The enumerator by number of parts for partitions with no part greater than $j$ is

$$F_j(t, u) = 1/(1 - ut)(1 - ut^2) \cdots (1 - ut^j)$$
$$= \Sigma p_j(t, k)u^k$$

Show that

$$(1 - ut)F_j(t, u) = (1 - ut^{j+1})F_j(t, ut)$$

and, hence, that

$$p_j(t, k) = t^k \frac{(1 - t^j)(1 - t^{j+1}) \cdots (1 - t^{j+k-1})}{(1 - t)(1 - t^2) \cdots (1 - t^k)}$$

is the enumerator for partitions with exactly $k$ parts and maximum part $j$.

6. In the notation of Problems 4 and 5, write

$$F(t, u) = \Sigma \Pi(t, k)u^k F_k(t, u)$$

Using the relations

$$(1 - ut)F(t, u) = F(t, ut)$$
$$(1 - ut)(F_k(t, u) + ut^{k+1}F_{k+1}(t, u)) = F_k(t, ut)$$

show that

$$(1 - t^k)\Pi(t, k) = t^{2k-1}\Pi(t, k - 1)$$

and by iteration that

$$\Pi(t, k) = \frac{t^{k^2}}{(1 - t)(1 - t^2) \cdots (1 - t^k)}$$

Derive the identity

$$\frac{1}{(1 - t)(1 - t^2)} \cdots = 1 + \frac{t}{(1 - t)^2} + \frac{t^4}{(1 - t)^2(1 - t^2)^2} + \cdots$$
$$+ \frac{t^{k^2}}{[(1 - t)(1 - t^2) \cdots (1 - t^k)]^k} + \cdots$$

7. The enumerator by number of parts for partitions with unequal parts and no part greater than $j$ is

$$G_j(t, a) = (1 + at)(1 + at^2) \cdots (1 + at^j)$$
$$= \Sigma u_j(t, k)a^k$$

Show that

$$(1 + at^{j+1})G_j(t, a) = (1 + at)G_j(t, at)$$

and, hence, that

$$u_j(t, k) = t^{\binom{k+1}{2}} \frac{(1 - t^j)(1 - t^{j-1}) \cdots (1 - t^{j-k+1})}{(1 - t)(1 - t^2) \cdots (1 - t^k)}, \qquad k < j$$

$$u_j(t, j) = t^{\binom{j+1}{2}}$$

8. If

$$I(t, a) = \frac{1}{1 - a} G(t, a) = \frac{1}{1 - a}(1 + at)(1 + at^2) \cdots = \Sigma U(t, k)a^k$$

where $G(t, a)$ is defined by (13), show that

$$(1 - a)I(t, a) = (1 - a^2t^2)I(t, at)$$
$$(1 - t^k)U(t, k) = U(t, k - 1) - t^kU(t, k - 2)$$

Verify the particular results

$$U(t, 0) = 1 \qquad (1-t)(1-t^2)U(t, 2) = 1 - t^2 + t^3$$
$$(1-t)U(t, 1) = 1 \qquad (1-t)(1-t^2)(1-t^3)U(t, 3) = 1 - t^2 + t^5$$

9. The enumerator by number of parts of partitions with odd unequal parts is

$$J(t, a) = (1 + at)(1 + at^3) \cdots (1 + at^{2n+1}) \cdots$$
$$= \Sigma v(t, k)a^k$$

Show that

$$J(t, a) = (1 + at)J(t, at^2)$$
$$(1 - t^{2k})v(t, k) = t^{2k-1}v(t, k-1)$$
$$(1-t^2)(1-t^4) \cdots (1-t^{2k})v(t, k) = t^{k^2}$$

in agreement with (18).

10. Show that

$$144[(1-t)(1-t^2)(1-t^3)(1-t^4)]^{-1} = 6(1-t)^{-4} + 18(1-t)^{-3}$$
$$+ 25(1-t)^{-2} + 16(1-t)^{-1}$$
$$+ 9(1-t^2)^{-1} + 18(1-t^2)^{-2}$$
$$+ 16(1-t^2)(1-t^3)^{-1} + 36(1-t^4)^{-1}$$

and, hence, that the denumerant $D(n; 1, 2, 3, 4)$ is given by

$$144D(n; 1, 2, 3, 4) = n^3 + 15n^2 + 63n + 65 + (27 + 18m)(1, 0) \text{ pcr } 2_n$$
$$+ 16(1, 0, -1) \text{ pcr } 3_n + 36(1, 0, 0, 0) \text{ pcr } 4_n$$

with $m = [n/2]$. With $< x >$ indicating the nearest integer to $x$, show that

$$<(12n + a + 5)^3/144> = D(12n + a; 1, 2, 3, 4), \qquad a \text{ even}$$
$$= D(12n + a; 1, 2, 3, 4) + n + 1, \qquad a \text{ odd}$$

11. The enumerator of compositions with exactly $m$ parts and no part greater than $s$, equation (35), is

$$c_m(t, s) = (t + t^2 + \cdots + t^s)^m$$

(*a*) Show that

$$(1-t)c_m(t, s) = t(1-t^s)c_{m-1}(t, s)$$
$$c_{m,n}(s) - c_{m,n-1}(s) = c_{m-1,n-1}(s) - c_{m-1,n-s-1}(s)$$

(*b*)
$$c_{m,n}(s) = \sum_0^m \binom{m}{k} c_{m-k,n-sk}(s-1)$$

12. The enumerator of compositions with no part greater than $s$ is [see equation (40)]

$$c(t, s) = \frac{t - t^{s+1}}{1 - 2t + t^{s+1}} = \frac{1-t}{1 - 2t + t^{s+1}} - 1$$
$$= \frac{t + t^2 + \cdots + t^s}{1 - t - t^2 - \cdots - t^s} = \frac{1}{1 - t - t^2 - \cdots - t^s} - 1$$
$$= \Sigma c_n(s)t^n$$

($a$) Show that the numbers $c_n(2)$ are Fibonacci numbers.

($b$) Derive the recurrence ($\delta_{ij}$ = Kronecker delta)

$$c_n(s) - 2c_{n-1}(s) + c_{n-s-1}(s) = \delta_{1n} - \delta_{s+1,n}$$

($c$) Verify the table for $c_n(s)$

| $s \backslash n$ | 1 | 2 | 3 | 4 | 5 | 6 | 7 | 8 | 9 | 10 |
|---|---|---|---|---|---|---|---|---|---|---|
| 1 | 1 | 1 | 1 | 1 | 1 | 1 | 1 | 1 | 1 | 1 |
| 2 | 1 | 2 | 3 | 5 | 8 | 13 | 21 | 34 | 55 | 89 |
| 3 | 1 | 2 | 4 | 7 | 13 | 24 | 44 | 81 | 149 | 274 |
| 4 | 1 | 2 | 4 | 8 | 15 | 29 | 56 | 108 | 208 | 401 |

13. ($a$) Show that the enumerator for compositions with no part greater than $s$ and at least one part equal to $s$ is

$$\begin{aligned} c^*(t, s) &= c(t, s) - c(t, s-1) \\ &= \frac{(1-t)^2 t^s}{(1 - 2t + t^s)(1 - 2t + t^{s+1})} \end{aligned}$$

($b$) Derive the relations

$$(1 - 2t + t^{s+1})c^*(t, s) = (t - 2t^2 + t^s)c^*(t, s-1)$$

$$c^*_n(s) - 2c^*_{n-1}(s) + c^*_{n-s-1}(s) = c^*_{n-1}(s-1) - 2c^*_{n-2}(s-1) + c^*_{n-s}(s-1)$$

($c$) Verify the table for $c^*_n(s)$

| $s \backslash n$ | 1 | 2 | 3 | 4 | 5 | 6 | 7 | 8 | 9 | 10 |
|---|---|---|---|---|---|---|---|---|---|---|
| 1 | 1 | 1 | 1 | 1 | 1 | 1 | 1 | 1 | 1 | 1 |
| 2 | | 1 | 2 | 4 | 7 | 12 | 20 | 33 | 54 | 88 |
| 3 | | | 1 | 2 | 5 | 11 | 23 | 47 | 94 | 185 |
| 4 | | | | 1 | 2 | 5 | 12 | 27 | 59 | 127 |

14. A tree coloring is made by placing one of $c$ colors independently on each of its points. If

$$q(x; c) = xq_1(c) + x^2 q_2(c) + \cdots$$

is the enumerator of these colored rooted trees by number of points, show that

$$q(x; c) = xc \exp\,(q(x; c) + q(x^2; c)/2 + \cdots + q(x^k; c)/k + \cdots)$$

From the relation, indicating a partial derivative by a suffix,

$$xq_x(x; c) = q(x; c)[1 + xq_x(x; c) + x^2 q_x(x^2; c) + \cdots]$$

determine the following

$$q_1(c) = c \qquad q_3(c) = 2c + 10\binom{c}{2} + 9\binom{c}{3}$$

$$q_2(c) = c + 2\binom{c}{2} \qquad q_4(c) = 4c + 44\binom{c}{2} + 102\binom{c}{3} + 64\binom{c}{4}$$

15. Show that the enumerator $u(x; c)$ of trees with colored points satisfies the relation

$$u(x; c) = q(x; c) - \tfrac{1}{2}q^2(x; c) + \tfrac{1}{2}q(x^2; c)$$

16. For rooted trees with colored points and no two adjacent points with the same color (point chromatic rooted trees, for short), show that the enumerator $p(x; c)$ satisfies

$$p(x; c) = xc \exp \frac{c-1}{c} (p(x; c) + p(x^2; c)/2 + \cdots + p(x^k; c)/k + \cdots)$$

$$xp_x(x; c) = p(x; c) \left[ 1 + \frac{c-1}{c} (xp_x(x; c) + x^2 p_x(x^2; c) + \cdots) \right]$$

Verify the following values

$$p_1 = c \qquad p_3 = 4\binom{c}{2} + 9\binom{c}{3}$$

$$p_2 = 2\binom{c}{2} \qquad p_4 = 8\binom{c}{2} + 54\binom{c}{3} + 64\binom{c}{4}$$

17. For point chromatic trees with enumerator $v(x; c)$, prove the relation

$$v(x; c) = p(x; c) - \frac{1}{2}\frac{c-1}{c} p^2(x; c)$$

18. For rooted trees with distinct labels on lines, show that the enumerator $r^*(x, y)$ is determined by

$$x(1 + y)r^*(x, y) = r(x, y)$$

where $r(x, y)$ is the enumerator for rooted trees with point labels [see equations (54) and (55)].

19. For rooted trees with line colors, show that the enumerator $q^*(x; c)$ is related to that for point colors (Problem 14) by

$$xcq^*(x; c) = q(x; c)$$

20. For line chromatic rooted trees, take $p^*(x; c)$ for the enumerator and $g(x; c) = 1 + xg_1(c) + x^2 g_2(c) + \cdots$ for the enumerator of planted trees (rooted trees with one line, the stem, added at the root) with a given color on the stem. Show that

$$p^*(x; c) = [1 + xg(x; c)]^c$$
$$g(x; c) = [1 + xg(x; c)]^{c-1}$$

and verify the particular values

$$p_0^* = 1 \qquad p_3^* = 4\binom{c}{2} + 16\binom{c}{3}$$

$$p_1^* = c \qquad p_4^* = 5\binom{c}{2} + 75\binom{c}{3} + 125\binom{c}{4}$$

$$p_2^* = 3\binom{c}{2} \qquad p_5^* = 6\binom{c}{2} + 279\binom{c}{3} + 1296\binom{c}{4} + 1296\binom{c}{5}$$

21. For trees with lines labeled, colored, and chromatic, find the relations

*Labeled*: $\quad x(1 + y)t^*(x, y) = t(x, y) + (y + y^2/2)r(x^2)$

*Colored*: $\quad xcu^*(x; c) = u(x; c) + \dfrac{c-1}{2} q(x^2; c)$

*Chromatic*: $\quad v^*(x; c) = p^*(x; c) - \frac{1}{2}xcg^2(x; c) + \frac{1}{2}xcg(x^2; c)$

22. For oriented rooted trees with point labels, show that the enumerator $\rho(x, y)$ satisfies $[\rho(x, 0) \equiv \rho(x)]$

$$\rho(x, y) = x(1 + y) \exp 2\,(\rho(x, y) + \rho(x^2)/2 + \cdots + \rho(x^k)/k + \cdots)$$
$$\rho(x, y) \exp 2\rho(x) = (1 + y)\rho(x) \exp 2\rho(x, y)$$
$$\alpha(x)\rho_x(x, y) = (1 + y)\rho_y(x, y) = \rho(x, y)[1 - 2\rho(x, y)]$$

with

$$\alpha(x) = \rho(x)/\rho'(x)(1 - 2\rho(x))$$

23. With $\tau(x, y)$ the enumerator for oriented trees with point labels [see equation (74)], show that (in the notation of the preceding problem)

$$\alpha(x)\tau'(x) = \rho(x)$$
$$\alpha(x)\tau_x(x, y) = (1 + y)\tau_y(x, y) = \rho(x, y)$$

24. Verify the following results for oriented trees. For the point colored case, the enumerators for rooted and free trees are given respectively by

$$\Pi(x; c) = xc \exp 2(\Pi(x; c) + \Pi(x^2; c)/2 + \cdots + \Pi(x^k; c)/k + \cdots)$$
$$v(x; c) = \Pi(x; c) - \Pi^2(x; c)$$

For the point chromatic case

$$\nu(x; c) = xc \exp 2\,\frac{c - 1}{c}\,(\nu(x; c) + \nu(x^2; c)/2 + \cdots)$$

$$\Phi(x; c) = \nu(x; c) - \frac{c - 1}{c}\,\nu^2(x; c)$$

For line labels

$$x(1 + y)\rho^*(x, y) = \rho(x, y)$$
$$x(1 + y)\tau^*(x, y) = \tau(x, y)$$

For line colors

$$xc\Pi^*(x; c) = \Pi(x; c)$$
$$xcv^*(x; c) = v(x; c)$$

For the line chromatic case

$$\nu^*(x; c) = p^*(2x; c)$$
$$\Phi^*(x; c) = v^*(2x; c)$$

25. *Oriented graphs.* These bear the same relation to the linear graphs of Section 11 that oriented trees bear to trees; each line may be marked with an arrow in one of two ways. Show that the store enumerator is $1 + 2x$ and that the cycle index is obtained from $C_n(t_1, t_2, \cdots, t_n)/n!$ by the rules:

$$t_{2m} \sim e_{2m}^{m-1}$$
$$t_{2m+1} \sim e_{2m+1}^{m}$$
$$t_i \circ t_j \sim e_M^D, \qquad M = [i, j], \qquad D = (i, j)$$

Note that cycles having both pairs $ij$ and $ji$ are impossible. Verify the particular cycle indexes

$$2E_2 = 1 + e_1$$
$$6E_3 = e_1^3 + 3e_2 + 2e_3$$
$$24E_4 = e_1^6 + 6e_1e_2^2 + 3e_2^2 + 8e_1e_3 + 6e_4$$

and, if the enumerator is

$$\theta_n(x) = E_n(1 + 2x, 1 + 2x^2, \cdots, 1 + 2x^n)$$

find the values

$$\theta_2(x) = 1 + x$$
$$\theta_3(x) = 1 + x + 3x^2 + 2x^3$$
$$\theta_4(x) = 1 + x + 4x^2 + 10x^5 + 4x^6 \qquad \text{(Harary, \textbf{14})}$$

26. Show that the connected graphs with one cycle of length $p$ and labels on points are enumerated by

$$D_p(x, y) = D_p(r(x, y), r(x^2), \cdots, r(x^p))$$

with $D_p(t_1, t_2, \cdots, t_p)$ the cycle indicator for the dihedral group [see equations (97) and (98)], and $r(x, y)$ the enumerator for rooted trees with point labels $[r(x, 0) = r(x)]$.

27. For the particular instance $p = 3$ of Problem 26, show that

$$\frac{\partial D_3(x, y)}{\partial y} = r_y(x, y)[r(x, y) + r(x^2) - t(x, y)]$$

and verify the numerical results

$$D_3(x, y) = x^3(1 + y + y^2/2 + y^3/6) + x^4(1 + 3y + 7y^2/2 + 12y^3/6 + 12y^4/24)$$
$$+ x^5(3 + 10y + 31y^2/2 + 81y^3/6 + 150y^4/24 + 150y^5/5!) + \cdots$$

28. (*a*) Show that the cycle index of oriented triangles is

$$t_1^3 + (t_1^3 + 2t_3)/3 = (4t_1^3 + 2t_3)/3$$

and, hence, that the enumerator of oriented connected graphs with one cycle of length 3 and distinct labels on points is

$$\delta_3(x, y) = (4\rho^3(x, y) + 2\rho(x^3))/3$$

(*b*) For an oriented square, find the similar results:

$$(4t_1^4 + 2t_2t_1^2 + t_2^2 + t_4)/2$$
$$\delta_4(x, y) = 2\rho^4(x, y) + \rho^2(x, y)\rho(x^2) + \tfrac{1}{2}[\rho^2(x^2) + \rho(x^4)]$$

(*c*) For an oriented pentagon, find

$$(16t_1^5 + 4t_5)/5$$
$$\delta_5(x, y) = \tfrac{4}{5}(4\rho^5(x, y) + \rho(x^5))$$

29. (*a*) For a connected graph with two cycles each of length 3 as in

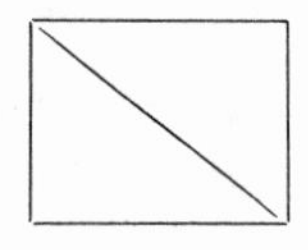

show that the cycle index is $(t_1^4 + 2t_1^2t_2 + t_2^2)/4$ and the enumerator with labels on points is

$$4E(x, y) = r^4(x, y) + 2r^2(x, y)r(x^2) + r^2(x^2)$$

(*b*) If this graph is oriented, show that there are 10 distinct orientations, that the cycle index is $8t_1^4 + 2t_1^2t_2$, and that

$$\varepsilon(x, y) = 8\rho^4(x, y) + 2\rho^2(x, y)\rho(x^2)$$

is the enumerator for point labels.

30. *Cacti.* These are formed from triangles as trees are formed from lines. Following Harary, let

$$\Delta(x) = 1 + x\Delta_1 + x^2\Delta_2 + \cdots$$

be the enumerator of rooted cacti by number of triangles, that is, $\Delta_n$ = number of cacti with $n$ triangles.

(*a*) Show that

$$\Delta(x) = \exp [s(x) + s(x^2)/2 + \cdots s(x^k)/k + \cdots]$$

with

$$s(x) = x(\Delta^2(x) + \Delta(x^2))/2 \qquad \text{(Harary and Uhlenbeck, \textbf{11})}$$

(*b*) With triangles individually labeled, and $\Delta(x, y)$ the enumerator by number of triangles, and number of triangles labeled, show that

$$\Delta(x, y) = \exp [s(x, y) + s(x^2)/2 + \cdots s(x^k)k + \cdots]$$

with

$$s(x, y) = x(1 + y)(\Delta^2(x, y) + \Delta(x^2))/2$$

(The functions $s(x)$ and $s(x, y)$ must not be confused with two-terminal series-parallel enumerators.)

(*c*) Combining (*a*) and (*b*) above, show that

$$\Delta(x, y) \exp s(x) = \Delta(x) \exp s(x, y)$$

or with $xy = z$ and

$$\begin{aligned} \Delta(x, z) &= \Gamma_0(z) + x\Gamma_1(z) + \cdots \\ \Delta(x, z) \exp s(x) &= \Delta(x) \exp s(x, z) \\ \Gamma_0(z) &= \exp z(1 + \Gamma_0(z))/2 \\ &= 1 + z + 3z^2/2 + 19z^3/b + 264z^4/4! + \cdots \end{aligned}$$

31. Show that the enumerator $\sigma(x; c)$ of two-terminal series-parallel networks with $c$ colors satisfies

$$1 + \sigma(x; c) = \exp [\alpha(x; c) + \alpha(x^2; c)/2 + \cdots + \alpha(x^k; c)/k + \cdots]$$

where $\alpha(x; c)$ is the enumerator of two-terminal essentially parallel networks; $\sigma(x; c) = 2\alpha(x; c) - xc$.

Writing

$$\sigma(x; c) = x\sigma_1(c) + x^2\sigma_2(c) + \cdots$$

verify the particular values

$$\sigma_1(c) = c \qquad \sigma_3(c) = 4c + 12\binom{c}{2} + 8\binom{c}{3}$$

$$\sigma_2(c) = 2c + 2\binom{c}{2} \qquad \sigma_4(c) = 10c + 60\binom{c}{2} + 102\binom{c}{3} + 52\binom{c}{4}$$

32. (*a*) If more than one line may join points of a linear graph, show that the store of objects has enumerator $(1 - x)^{-1}$, while the cycle index remains unchanged, so that the enumerator by number of lines of graphs with $n$ points is

$$\lambda_n(x) = G_n[(1 - x)^{-1}, (1 - x^2)^{-1}, \cdots, (1 - x^n)^{-1}]$$

with $G_n$ as in equation (90).

(*b*) Verify the particular values

$$\lambda_1(x) = 1$$
$$\lambda_2(x) = (1 - x)^{-1}$$
$$\lambda_3(x) = D(x; 123)$$

with $D(x; 123)$ the denumerant of partitions into parts 1, 2, and 3. Hence $\lambda_{3k}$, the number of graphs with 3 points and $k$ lines, with lines in parallel allowed, is the nearest integer to $(k + 3)^2/12$.

33. Write the enumerator $L(x, y)$ of linear graphs without lines in parallel by number of points and lines as

$$\begin{aligned} 1 + L(x, y) &= (1 - y)^{-1}h(x, y) \\ &= (1 - y)^{-1}[1 + xh_1(y) + x^2h_2(y) + \cdots] \end{aligned}$$

and the corresponding enumerator of connected graphs as

$$C(x, y) = c_0(y) + xc_1(y) + x^2c_2(y) + \cdots$$

Show that

$$h_x(x, y) = h(x, y)(C_x(x, y) + xC_x(x^2, y^2) + \cdots + x^{k-1}C_x(x^k, y^k) + \cdots)$$

and

$$nh_n(y) = C_n(y)h_0(y) + C_{n-1}(y)h_1(y) + \cdots + C_1(y)h_{n-1}(y)$$

with

$$C_1(y) = c_1(y) = y^2 \qquad C_3(y) = 3c_3(y) + c_1(y^3)$$
$$C_2(y) = 2c_2(y) + c_1(y^2) \qquad C_4(y) = 4c_4(y) + 2c_2(y^2) + c_1(y^4)$$

and so on.

From the values

$$c_1(y) = y^2 \qquad c_3(y) = y^3 + 2y^4$$
$$c_2(y) = y^3 \qquad c_4(y) = 2y^4 + 3y^5$$

determine $(h_0(y) = 1)$

$$h_1(y) = y^2 \qquad h_3(y) = y^3 + 2y^4 + y^5 + y^6$$
$$h_2(y) = y^3 + y^4 \qquad h_4(y) = 2y^4 + 4y^5 + 3y^6 + y^7 + y^8$$

Compare Table 7.

34. *Chemical "trees"*. These are particular trees appearing in the structural diagrams of chemical compounds. A number of enumerations of this kind appear in Pólya, **20**; indeed, they supply the greater part of the early interest in the study of trees. For brevity, consider only the paraffins. These are rooted trees with two kinds of points other than the root: outer points with but one line, and inner points each with four lines; or, in chemical language, H atoms and C atoms respectively.

(*a*) For a paraffin tree with $p$ points, $p_1$ outer and $p_4$ inner and $s = p - 1$ lines, show that

$$p_1 + p_4 = p$$
$$p_1 + 4p_4 = 2s = 2p - 2$$

so that

$$p_1 = 2p_4 + 2$$

(*b*) If $p(x) = 1 + xp_1 + x^2p_2 + \cdots$ is the enumerator of such rooted trees by number of inner points, show that

$$p(x) = 1 + x(p^3(x) + 3p(x)p(x^2) + 2p(x^3))/6$$

35. If $a(x) = a_1x + a_2x^2 + \cdots$, and

$$a^m(x) = a_1(m)x^m + a_2(m)x^{m+1} + \cdots + a_n(m)x^{n+m-1} + \cdots$$

show that

$$a_1(m) = a_1^m$$
$$a_2(m) = ma_2a_1^{m-1}$$
$$a_3(m) = ma_3a_1^{m-1} + \binom{m}{2} a_2^2a_1^{m-2}$$
$$a_4(m) = ma_4a_1^{m-1} + \binom{m}{2} 2a_3a_2a_1^{m-2} + \binom{m}{3} a_2^3a_1^{m-3}$$

and, for $n > 0$,

$$a_{n+1}(m) = Y_n(fa_2, f2a_3, \cdots, fn!\, a_{n+1})/n!, \qquad f^j \equiv f_j = (m)_j a_1^{m-j}$$

36. (*a*) Using the results of Problem 35, and $a(x) = r(x)$, the enumerator of rooted trees, find the results (see Table 8)

$$r_1(m) = 1$$
$$r_2(m) = m$$
$$r_3(m) = 2m + \binom{m}{2}$$
$$r_4(m) = 4m + 4\binom{m}{2} + \binom{m}{3}$$
$$r_5(m) = 9m + 12\binom{m}{2} + 6\binom{m}{3} + \binom{m}{4}$$
$$r_6(m) = 20m + 34\binom{m}{2} + 24\binom{m}{3} + 8\binom{m}{4} + \binom{m}{5}$$

(*b*) If $r_n(m) = \Sigma b_{n,k}\binom{m}{k}$, show that

$$b_{n,n-1} = 1$$
$$b_{n,n-2} = 2(n-2)$$
$$b_{n,n-3} = 4\binom{n-2}{2}$$
$$b_{n,n-4} = n - 4 + 8\binom{n-2}{3}$$
$$b_{n,n-5} = 4\binom{n-4}{2} + 16\binom{n-2}{4}$$
$$b_{n,n-6} = 4(n-6) + 12\binom{n-4}{3} + 32\binom{n-2}{5}$$

37. *Necklaces.*

(*a*) Show that the enumerator for necklaces with $n$ beads, each of which may have any of $c$ colors, when equivalences produced by turns in one direction only are considered, is

$$N_1(x_1, x_2, \cdots, x_c) = \frac{1}{n} \sum_{d|n} \phi(d)(x_1^d + \cdots + x_c^d)^{n/d}$$

Show similarly that two-sided necklaces (considering equivalences produced by turns in both directions) have the enumerator

$$N_2(x_1, x_2, \cdots, x_c) = D_n(s_1, s_2, \cdots, s_n)$$

where $D_n$ is the cycle index of the dihedral group [equations (97) and (98)], and $s_i = x_1^i + x_2^i + \cdots + x_c^i$.

(*b*) Using the first of the results in (*a*), show that the number of circular permutations of $n$ objects of $c$ kinds, with repetitions allowed, is given by

$$N_1(n) = \frac{1}{n} \sum_{d|n} \phi(d) c^{n/d}$$

and that the number of circular permutations of $n$ objects of specification $(n_1, n_2, \cdots, n_m)$ is given by

$$N_1(n_1, n_2, \cdots, n_m) = \frac{1}{n} \Sigma \phi(d) \frac{(n/d)!}{(n_1/d)! \cdots (n_m/d)!}$$

where $d$ is a divisor of the greatest common divisor of $n_1, n_2, \cdots, n_m$.* Similar but more complicated results may be obtained from the second enumerator.

* According to Dickson, **4**, vol. I, the first of these has been given by P. A. MacMahon, *Proc. London Math. Soc.*, vol. 23 (1891–2), pp. 305–313, while the second is quoted by E. Lucas, *Théorie des nombres*, Paris, 1891, pp. 501–503 from C. Moreau.

CHAPTER 7

# Permutations with Restricted Position I

## 1. INTRODUCTION

This chapter is devoted to the enumeration of permutations satisfying prescribed sets of restrictions on the positions of the elements permuted. The permutations appearing in the problème des rencontres, described and solved in Chapter 3, provide the simplest example of a restricted position problem in which each element has some restriction (element $i$ is forbidden position $i$). A related problem is that called by E. Lucas, **11**, the "problème des ménages"; this asks for the number of ways of seating $n$ married couples at a circular table, men and women in alternate positions, so that no wife is next to her husband. Seating the wives first, which may be done in $2n!$ ways, each husband is excluded from the two seats beside his wife, and the number of ways of seating the husbands is independent of the seating arrangement of the wives; its enumeration is known as the reduced ménage problem. Numbering the husbands 1 to $n$, the reduced problem may be taken as the enumeration of permutations such that $i$ may not be in positions $i$ or $i + 1$, $i = 1$ to $n - 1$, and $n$ may not be in positions $n$ or 1 (because the table is circular).

Any restrictions of position may be represented on a square, with the elements to be permuted as column heads and the positions as row heads, by putting a cross at a row-column intersection to mark a restriction. For example, for permutations of four (distinct) elements, the arrays of restrictions for the rencontres and reduced ménage problems mentioned above are

|   | 1 | 2 | 3 | 4 |
|---|---|---|---|---|
| 1 | × |   |   |   |
| 2 |   | × |   |   |
| 3 |   |   | × |   |
| 4 |   |   |   | × |

rencontres

|   | 1 | 2 | 3 | 4 |
|---|---|---|---|---|
| 1 | × |   |   | × |
| 2 | × | × |   |   |
| 3 |   | × | × |   |
| 4 |   |   | × | × |

ménages

Since a square of side $n$ has $n^2$ cells, and a cross may or may not appear in each cell, it is clear that with $n$ elements $2^{n^2}$ problems are possible (this includes permutations without restriction, for which no cell has a cross). However, many of these are not distinct since, from the enumeration standpoint, the relative rather than the absolute position of the crosses is important; for example, all $n^2$ problems having just one cross on the board are alike. The exact number of distinct problems, for any $n$, is not known, but some progress in this direction will appear in this chapter. Moreover, many of the distinct problems are inconsequential. Those of most interest seem to involve arrays of the following kinds: (i) "diagonal" or "staircase" like those of rencontres and ménages above; (ii) rectangles; and (iii) triangles.

The "staircase" arrays are related to "Latin rectangles". A Latin rectangle is a rectangular array of elements, each row of which is a permutation of (the same) $n$ elements, and each column of which is a set of distinct elements. Latin rectangles were first studied by Euler in 1782, and the term Latin merely indicates that the elements are Latin letters (it distinguished these arrays from Graeco-Latin rectangles—also considered by Euler—in which each cell of the array contained both a Greek and a Latin letter). A Latin rectangle with two rows and $n$ columns may be normalized by putting the first row in standard order; then the second row may contain only the permutations enumerated for the problème des rencontres. The number of two-line Latin rectangles, say $L(2, n)$, is then $n!\, D_n$, with $D_n$ the displacement number of Chapter 3. A more complicated relationship exists between the number of three-line Latin rectangles and the ménage numbers; it will be given in the next chapter. The next case in this line, a sort of three-ply staircase, is much harder and has a devious relation to four-line Latin rectangles, about which, in fact, little is known.

Rectangular arrays appear when some elements are alike as in problems of card matching, such as those in studies of extrasensory perception. Finally, the triangular arrays are closely associated with Simon Newcomb's problem, which is as follows: a deck of cards of arbitrary specification is dealt out into a single pile so long as the cards are in rising order, with like cards counted as rising; with the occurrence of a card less than its predecessor, a new pile is started. In how many ways do $k$ piles appear? This problem also will be treated in the next chapter.

## 2. THE PROBLEM OF THE ROOKS

A rook is a chessboard piece which takes on rows and columns. The problem of the rooks is that of enumerating the number of ways of putting $k$ non-taking rooks on a given chessboard, a chessboard being an arbitrary array of cells arranged in rows and columns like the arrays considered above. Stated otherwise, the problem is that of putting $k$ rooks on a chessboard in such a way that no two rooks are in the same row or column.

Any problem of permutations with restricted position may be reduced to a rook problem on the board corresponding to the restricted positions, that is, the board made up of the cells with crosses in the array already mentioned.

In fact, if $r_k$ is the number of ways of putting $k$ non-taking rooks on this board, and if $N_j$ is the number of permutations of $n$ distinct elements with exactly $j$ elements in restricted positions, then

$$N_n(t) = \Sigma N_j t^j = \Sigma r_k (n-k)!\,(t-1)^k \tag{1}$$

This follows from the principle of inclusion and exclusion. By equation (3.5$a$)

$$N_n(t) = \Sigma NS_k (t-1)^k$$

where $NS_k = \Sigma N(a_{i_1} a_{i_2} \cdots a_{i_k})$. In the present application, the properties $a_1, a_2, \cdots$ are defined by the restricted positions, that is, each $a_i$ is the property that some given element $j$ be in a given (restricted) position $k$. The total number of properties is equal to the number of restrictions imposed, that is, the number of crosses on the corresponding array. Some of these properties are compatible, some are not; if $i$ is in position $j$, then no other element may be in the same position and no other position is possible for $i$. These are just the conditions which determine the numbers $r_k$. Hence, the number of terms in the sum $NS_k$ is $r_k$, and for any $k$ compatible properties, there are $(n-k)!$ positions for the remaining elements. Hence

$$NS_k = r_k (n-k)!$$

as in (1).

Introducing the ordinary generating function $R(x)$ of the numbers $r_k$, that is,

$$R(x) = \Sigma r_k x^k \tag{2}$$

it is clear that $N_n(t)$ is determined by the substitution $x^k = (n-k)!\,(t-1)^k$; writing $(n-k)! = E^{-k} n!$, equation (1) may be rewritten as the symbolic equation

$$N_n(t) = R[(t-1)E^{-1}]n! \tag{3}$$

The generating function $R(x)$ will be called a rook polynomial. The generating function $N_n(t)$ will be called a *hit* polynomial; a hit is the appearance of an element of a permutation in a position forbidden to it, and the hit polynomial enumerates permutations by number of hits.

It is convenient to have a symbolic equation operating on 0!, rather than $n!$ as in (3); by equation (1)

$$N_n(t) = (1 - t)^n r(E(1 - t)^{-1})0! \tag{3a}$$

if

$$r(x) = \Sigma(-1)^k r_k x^{n-k} = x^n R(-x^{-1}) \tag{2a}$$

The polynomial $r(x)$ is called the *associated* rook polynomial.

These are basic results for this chapter and the next. To emphasize their importance, they are summarized in

**Theorem 1.** *For any problem of permutations with restricted position, the rook polynomial, $R(x)$, of the board defined by the given restrictions or its associated polynomial, $r(x)$, determine the hit polynomial $N_n(t)$ by*

$$N_n(t) = \Sigma r_k(n - k)!\,(t - 1)^k \tag{1}$$

$$= R[(t - 1)E^{-1}]n! \tag{3}$$

$$= (1 - t)^n r[E(1 - t)^{-1}]0! \tag{3a}$$

It is worth noting here that the rook polynomial is effectively a factorial moment generating function for the probability distribution $P_n(t) = N_n(t)/n!$. Indeed, by equation (3.5)

$$N_n(t) = n!\,P_n(t) = n!\,\Sigma(m)_k(t - 1)^k/k!$$

with $(m)_k$ the $k$th factorial moment; comparing this with (1) leads to

$$r_k = \binom{n}{k}(m)_k$$

$$R(x) = (1 + (m)x)^n, \qquad (m)^k \equiv (m)_k \tag{2b}$$

$$r(x) = (x - (m))^n, \qquad (m)^k \equiv (m)_k \tag{2c}$$

An immediate convenience of the associated rook polynomial is its use in expressing the hit polynomials in terms of derivatives. In the first instance, with a prime denoting a derivative,

$$\begin{aligned}(1 - t)^{n-1}r'[E(1 - t)^{-1}]0! &= \sum_0^{n-1} r_k(n - k)!\,(t - 1)^k \\ &= N_n(t) - r_n(t - 1)^n \\ &= N_n(t) - (1 - t)^n r(0) \qquad (4)\end{aligned}$$

More generally, with $r^{(k)}(x)$ the $k$th derivative,

$$\begin{aligned}(1 - t)^{n-k}r^{(k)}[E(1 - t)^{-1}]0! &= N_n(t) - (1 - t)^n r(0) - (1 - t)^{n-1}r^{(1)}(0) \\ &\quad - \cdots - (1 - t)^{n-k+1}r^{(k-1)}(0) \qquad (5)\end{aligned}$$

With these relations, any recurrence relation in $r(x) \equiv r_n(x)$ and its derivatives is immediately transcribed into a recurrence relation for the hit polynomials; note that $r_n^{(k)}(0)$ is a number. Thus, in the simple instance where

$$\Sigma a_k r_{n-k}(x) = \Sigma b_k r'_{n-k}(x)$$

with $a_k$ and $b_k$ numbers, it follows at once that

$$\Sigma a_k(1-t)^k N_{n-k}(t) = \Sigma b_k(1-t)^{k+1} N_{n-k}(t) - (1-t)^{n+1}\Sigma b_k r_{n-k}(0)$$

Relations of this kind, for the numbers $N_{n-k}(0)$, appear in Yamamoto, **15.**

It should be noticed that for the use of any of these results it is not necessary that elements permuted be distinct. But like elements are forbidden the same positions and the array of restrictions is then one of rectangular blocks. From the standpoint of the array, the labeling of the elements is immaterial, and the hit polynomial $N_n(t)$ for like elements differs from that for unlike elements of the same total and with the same array only by a constant factor.

*Example 1.* For the problème des rencontres, the restrictions are: Element $i$ may not be in position $i$; the chessboard consists of $n$ disjunct (no two in the same row or column) cells, and the rook polynomial is $(1 + x)^n$, since $r_k$ is simply the number of ways of choosing $k$ of $n$ cells, which is the binomial coefficient $\binom{n}{k}$. [As a verification of equation (1), note that, by equation (3.15), $NS_k = \binom{n}{k}(n-k)!$]. In the notation of Chapter 3, $N_n(t) = D_n(t)$, and by (3)

$$\begin{aligned} D_n(t) &= [1 + (t-1)E^{-1}]^n n! \\ &= (E - 1 + t)^n 0! = (\Delta + t)^n 0! = \Sigma \binom{n}{k} \Delta^{n-k}0!\, t^k \\ &= (D + t)^n, \qquad D^k \equiv D_k = \Delta^k 0! \end{aligned}$$

in agreement with equation (3.18).

As an illustration of the use of equation (4), note that

$$r'_n(x) = n(x-1)^{n-1} = nr_{n-1}(x)$$

so that, because $r_n(0) = (-1)^n$,

$$D_n(t) = nD_{n-1}(t) + (t-1)^n$$

which is equation (3.20).

*Example 2.* Incomplete rencontres. Suppose only $m$ of the $n$ elements have the restrictions of the preceding example, the remaining being unrestricted. Then the chessboard has $m$ disjunct cells, the rook polynomial is $R(x) = (1 + x)^m$, and the associated rook polynomial is $r(x) \equiv r_n(x, n - m) = x^{n-m}(x-1)^m$. Note that $r_n(x, 0) = (x-1)^n$, this being the corresponding polynomial of the preceding example, which explains the choice of notation. Write $D_n(t, n-m)$ for the hit polynomial; then by Theorem 1,

$$\begin{aligned} D_n(t, n-m) &= \Sigma \binom{m}{k} (n-k)!\,(t-1)^k \\ &= [1+(t-1)E^{-1}]^m n! = (\Delta + t)^m (n-m)! \\ &= (1-t)^n (E(1-t)^{-1})^{n-m}(E(1-t)^{-1}-1)^m 0! \end{aligned}$$

Simplifying the last result, it appears that

$$\begin{aligned} D_n(t, n-m) &= E^{n-m}(E-1+t)^m 0! \\ &= (\Delta+1)^{n-m}(\Delta+t)^m 0! \\ &= (D+1)^{n-m}(D+t)^m, \qquad D^k \equiv D_k = \Delta^k 0! \end{aligned}$$

Using the binomial recurrence in the first leads to

$$D_n(t, n-m) = D_n(t, n-m+1) + (t-1)D_{n-1}(t, n-m)$$

while by equation (4), $r_n(0, n-m) = 0$, $n > m$, and

$$r'_n(x, n-m) = (n-m)r_{n-1}(x, n-1-m) + mr_{n-1}(x, n-m)$$
$$D_n(t, n-m) = (n-m)D_{n-1}(t, n-1-m) + mD_{n-1}(t, n-m), \qquad n > m$$

The first few hit polynomials are as follows

| $n-m, n$ | 1 | 2 | 3 |
|---|---|---|---|
| 0 | $t$ | $1+t^2$ | $2+3t+t^3$ |
| 1 | 1 | $1+t$ | $3+2t+t^2$ |
| 2 | | 2 | $4+2t$ |
| 3 | | | 6 |

## 3. PROPERTIES OF ROOK POLYNOMIALS

For a chessboard $C$ with disjunct parts $C_1$ and $C_2$, that is, with no cells of $C_1$ in the same row or column as those of $C_2$, it is clear that

$$R(x, C) = R(x, C_1)R(x, C_2) \tag{6}$$

This result has already appeared implicitly in Examples 1 and 2. Hence, the basic rook polynomials are those of connected boards.

Next, for any board, an expansion with respect to any cell follows from the remark that the $r_k$ arrangements of rooks may be divided into those with a rook in a given cell and those without a rook in this cell. With a rook in a given cell, no other rook may be in a cell in the same row or column; hence, these numbers are enumerated on a board with this row and column removed, which may be indicated by $C_i$ ($i$ to indicate inclusion of the given cell). In the contrary case, only the given cell is removed; the board is $C_e$ ($e$ for exclusion). Hence

$$r_k(C) = r_{k-1}(C_i) + r_k(C_e)$$

and

$$R(x, C) = xR(x, C_i) + R(x, C_e) \tag{7}$$

*Example 3.* (*a*) For the triangular board

$$\begin{matrix} \times & \times \\ & \times \end{matrix}$$

expansion with respect to the cell in the first row and column determines the derived boards

$$C_i = \times, \qquad C_e = \begin{matrix} \times \\ \times \end{matrix}$$

and

$$R(x) = x(1 + x) + 1 + 2x = 1 + 3x + x^2$$

If brackets are used to indicate the polynomials of the boards they enclose, this may be abbreviated in a natural way by

$$\begin{bmatrix} \times & \times \\ & \times \end{bmatrix} = \begin{bmatrix} \times \\ \times \end{bmatrix} + x[\times]$$

(*b*) In the same way

$$\begin{bmatrix} \times & \times & \\ & \times & \times \end{bmatrix} = \begin{bmatrix} \times & \\ \times & \times \end{bmatrix} + x[\times \times]$$

$$= 1 + 3x + x^2 + x(1 + 2x) = 1 + 4x + 3x^2$$

These examples suggest two things: first, the expansion may be used to determine polynomials by recurrence and, second, with a library of classes of such polynomials a single application may be enough to determine completely a polynomial of a given array.

If one or both boards appearing on expansion with respect to a single cell do not have known polynomials, expansion may be continued as often as desired, and cells may be picked at pleasure in each of the boards which appear, until boards with known polynomials are obtained. Alternatively, a given board may be expanded with respect to several given cells; this is equivalent to picking the same cells for expansion in each of the subboards of the iterative process suggested above. The formal expression for expansion with respect to any number of cells numbered 1, 2, $\cdots$ is as follows:

$$R = R[(e_1 + xi_1)(e_2 + xi_2) \cdots] \tag{8}$$

This is a symbolic expression like equation (3.1*a*) and has the same rules of interpretation; the variable $x$ of the polynomial $R(x)$ is carried silently. Thus,

$$R(e_1 + xi_1) = R(e_1) + xR(i_1)$$

is an abbreviation of equation (7);

$$R[(e_1 + xi_1)(e_2 + xi_2)] = R(e_1e_2) + xR(e_1i_2) + xR(e_2i_1) + x^2R(i_1i_2)$$

expresses the partition of the rook arrangements into those with no rooks on cells 1 and 2, $R(e_1e_2)$, with none on cell 1 and one on cell 2, $xR(e_1i_2)$, and so on.

With $n$ cells for expansion, there are $2^n$ terms in the development of (8), exhausting all possibilities.

If the cells for expansion are all on a single row or column, it is impossible to include two or more, and equation (8) reduces to

$$R = R(e_1e_2\cdots) + x[R(i_1e_2e_3\cdots) + R(e_1i_2e_3\cdots) + R(e_1e_2i_3\cdots) + \cdots] \tag{8a}$$

There are as many terms within the square brackets as there are cells for expansion, and each term contains a single inclusion symbol.

## 4. RECTANGULAR BOARDS

Take the rectangle as having $m$ rows and $n$ columns. On expanding with respect to all the $n$ cells in one row, then by (8a) if $R_{m,n}(x)$ is the rook polynomial,

$$R_{m,n}(x) = R_{m-1,n}(x) + xnR_{m-1,n-1}(x) \tag{9}$$

since $R(e_1e_2\cdots e_n)$ is the rectangle with one row wiped out, and for any $j$, $R(e_1\cdots e_{j-1}i_je_{j+1}\cdots e_n)$ is the rectangle without one row and one column. In the same way, or by symmetry,

$$R_{m,n}(x) = R_{m,n-1}(x) + xmR_{m-1,n-1}(x) \tag{9a}$$

It is immediately evident that

$$R_{1,n}(x) = 1 + nx$$

and by (9)

$$R_{2,n}(x) = 1 + 2nx + n(n-1)x^2$$
$$R_{3,n}(x) = 1 + 3nx + 3n(n-1)x^2 + n(n-1)(n-2)x^3$$

These suggest that

$$R_{m,n}(x) = 1 + mnx + \binom{m}{2}(n)_2x^2 + \cdots + \binom{m}{k}(n)_kx^k + \cdots + (n)_mx^m \tag{10}$$

for $m \leq n$, and this is verified by the use of (9) and mathematical induction, (since it is true for $m = 1$).

It is verified more simply by direct reasoning, as follows: $k$ non-taking rooks require a $k$ by $k$ board, since no two may be in the same row or column, and can be put on a given $k$ by $k$ board in $k!$ ways. Also a

$k$ by $k$ board can be obtained by choosing rows in $\binom{m}{k}$ ways, and columns in $\binom{n}{k}$ ways: Hence,

$$r_k = k!\binom{m}{k}\binom{n}{k} = \binom{m}{k}(n)_k$$

as in (10).

Another representation of the polynomials is worth noting. Laguerre polynomials may be defined by*

$$n!\,L_n^\alpha(x) = e^x x^{-\alpha} D^n(e^{-x}x^{n+\alpha}), \qquad D = d/dx$$

so that

$$L_0^\alpha(x) = 1$$

$$L_1^\alpha(x) = \alpha + 1 - x$$

$$L_n^\alpha(x) = \sum_{k=0}^{n}\binom{n+\alpha}{n-k}\frac{(-x)^k}{k!}$$

Hence, for integral $\alpha$,

$$R_{n,n+\alpha}(x) = n!\,x^n L_n^\alpha(-x^{-1}) \tag{11}$$

If an $m$ by $n$ rectangle is an array of forbidden positions for permutations of $n$ elements, hence $m \leq n$, it follows from (1) that the hit polynomial, say $N_n(t)$, is given by

$$N_n(t) = \Sigma N_{n,j}t^j = \Sigma\binom{m}{k}\binom{n}{k}k!\,(n-k)!\,(t-1)^k = n!\,t^m \tag{12}$$

That is, all permutations have exactly $m$ elements in restricted positions.

Because of (12) the rectangular polynomials occupy a central position in the theory of permutations with restricted position. If $R(x)$ is the rook polynomial for an arbitrary board associated with restrictions on positions of $n$ elements, and $A_n(t)$ is the corresponding hit polynomial, by (3)

$$A_n(t) = \Sigma A_{n,j}t^j = R[(t-1)E^{-1}]n!$$

and by (12)

$$n!\,t^j = R_{n,j}[(t-1)E^{-1}]n!$$

where $R_{n,j}(x)$ is the rook polynomial of an $n$ by $j$ rectangle. Hence,

$$n!\,R[(t-1)E^{-1}]n! = \Sigma A_{nj}R_{n,j}[(t-1)E^{-1}]n!$$

which is equivalent to

$$n!\,R(x) = \Sigma A_{nj}R_{n,j}(x) \tag{13}$$

This result may be used in many ways; for example, to develop recurrence relations or to find curious identities, as appears below.

* A. Erdelyi, et al., *Higher Transcendental Functions*, vol. 2, p. 188, New York, 1953.

*Example 4.* For the rencontres problem (Example 1), the rook polynomial is $(1 + x)^n$, the hit polynomial $D_n(t) = (D + t)^n$. Hence by (13)

$$n!\,(1 + x)^n = \Sigma \binom{n}{j} D_{n-j} R_{n,j}(x)$$

a curious relation, having the instances

$$\begin{aligned} 2(1 + x)^2 &= R_{20} + R_{22} = 1 + (1 + 4x + 2x^2) \\ 6(1 + x)^3 &= 2R_{30} + 3R_{31} + R_{33} \\ &= 2 + 3(1 + 3x) + (1 + 9x + 18x^2 + 6x^3) \end{aligned}$$

*Example 5.* (*a*) Since $(d/dx)(1 + x)^n = n(1 + x)^{n-1}$, it follows from Example 4 that

$$\Sigma \binom{n}{m} D_{n-m} \frac{d}{dx} R_{n,m}(x) = n^2 \Sigma \binom{n-1}{m} D_{n-1-m} R_{n-1,m}(x)$$

Hence

$$(d/dx) R_{n,m}(x) = nm R_{n-1,m-1}(x)$$

which is verified at once from (10).

(*b*) Write the relation of Example 4 in symbolic form as

$$n!\,(1 + x)^n = (D + R_{n,.})^n, \qquad D^k \equiv D_k, \qquad R^k_{n,.} \equiv R_{n,k}(x)$$

Since $R_{n,k}(0) = 1$, it follows that $n! = (D + 1)^n$, as has already appeared in Chapter 3. If the equation above is rewritten,

$$n!\,(1 + x)^n = (D + 1 + R_{n,.} - 1)^n$$

or

$$\Sigma \binom{n}{m} n!\,x^m = \Sigma \binom{n}{m} (n - m)!\,(R_{n,.} - 1)^m$$

The result just found suggests that

$$(n)_m x^m = (R_{n,.} - 1)^m$$

and, in fact,

$$\begin{aligned} nx &= R_{n,1} - R_{n,0} \\ n(n - 1)x^2 &= R_{n,2} - 2R_{n,1} + R_{n,0} \end{aligned}$$

A proof may be obtained by mathematical induction and the relation

$$\begin{aligned} (R_{n,.} - 1)^{m+1} &= \Sigma \binom{m+1}{k} (-1)^k R_{n,m+1-k} \\ &= \Sigma \binom{m}{k} (-1)^k [R_{n,m-k} + nxR_{n-1,m-k}] \\ &\quad + \Sigma \binom{m}{k-1} (-1)^k R_{n,m+1-k} \\ &= nx\Sigma \binom{m}{k} (-1)^k R_{n-1,m-k} = nx(R_{(n-1),.} - 1)^m \end{aligned}$$

Note that

$$\begin{aligned} (n)_m x^m &= (R_{n,.} - 1)^m, \qquad R^k_n \equiv R_{n,k}(x), \qquad m \le n \\ R_{n,m}(x) &= (1 + (n)x)^m, \qquad (n)^k \equiv (n)_k, \qquad m \le n \end{aligned}$$

are inverse relations, the latter being a version of (10).

(*c*) Write, as in Chapter 3, $D_{n,m} = \binom{n}{m} D_{n-m}$. Then,

$$n!\,(1 + x)^n = n(1 + x)\,[(n - 1)!\,(1 + x)^{n-1}]$$

will imply a recurrence relation for $D_{n,m}$ if $n(1 + x)R_{n-1,m}$ may be expressed in terms of rectangular polynomials all of which have one index equal to $n$. The relations required, taken from (9) and (9*a*), are

$$R_{n,m+1} - R_{n,m} = nxR_{n-1,m}$$
$$(n - m)R_{n,m} + mR_{n,m-1} = nR_{n-1,m}$$

Hence,

$$R_{n,m+1} + (n - m - 1)R_{n,m} + mR_{n,m-1} = n(1 + x)R_{n-1,m}$$

and

$$\Sigma D_{n,m}R_{n,m} = \Sigma D_{n-1,m}[R_{n,m+1} + (n - m - 1)R_{n,m} + mR_{n,m-1}]$$

Finally,

$$D_{n,m} = D_{n-1,m-1} + (n - m - 1)D_{n-1,m} + (m + 1)D_{n-1,m+1}$$

Notice that for $m = 0$

$$\begin{aligned} D_{n,0} &= (n - 1)D_{n-1,0} + D_{n-1,1} \\ &= (n - 1)(D_{n-1,0} + D_{n-2,0}) \end{aligned}$$

which is Euler's relation, Equation 3.24. Also notice that the recurrence implies

$$\begin{aligned} D_n(t) &= \Sigma D_{n,m}t^m \\ &= (n - 1 + t)D_{n-1}(t) + (1 - t)D'_{n-1}(t) \end{aligned}$$

the prime as usual indicating differentiation. For consistency with the result of Problem 3.6, this requires $D'_n(t) = nD_{n-1}(t)$, which is Equation (3.19).

The relations of rectangular polynomials appearing in Example 5 may be systematized by introducing the operators $E_n$, $E_m$ defined by

$$E_nR_{n,m}(x) = R_{n+1,m}(x)$$
$$E_mR_{n,m}(x) = R_{n,m+1}(x)$$

The basic recurrences in equations (9) and (9*a*) are equivalent to

$$[E_mE_n - E_n - (n + 1)x]R_{m,n} = 0 \tag{14}$$
$$[E_mE_n - E_m - (m + 1)x]R_{m,n} = 0$$

The operators $E_m$ and $E_n$ work only on the suffixes of $R_{m,n}$; they are independent and commutable. Since $R_{m,n} = R_{n,m}$, the two forms are really the same. Rewriting the first as

$$(E_m - 1)R_{m,n} = nxR_{m,n-1}$$

it follows that

$$(E_m - 1)^2R_{m,n} = nx(E_m - 1)R_{m,n-1} = n(n - 1)x^2R_{m,n-2}$$

Hence

$$(E_m - 1)^kR_{m,n} = (n)_kx^kR_{m,n-k} \tag{15}$$

This is slightly more general than the relation

$$(n)_m x^m = (R_{n,\cdot} - 1)^m = (E_m - 1)^m R_{n,0}$$

appearing in Example 5(*b*).

Combination of the two equations of (14) gives another relation used in Example 5, namely,

$$[(n - m)E_m E_n + mE_n - nE_m]R_{m-1,n-1} = 0 \tag{16}$$

or

$$[n - m + mE_m^{-1} - nE_n^{-1}]R_{m,n} = 0$$

Hence

$$[n - m + mE_m^{-1}]R_{m,n} = nE_n^{-1}R_{m,n}$$

$$[(n - m)(n - m - 1) + 2m(n - m)E_n^{-1} + m(m - 1)E_m^{-2}]R_{m,n} = n(n - 1)E_n^{-2}R_{m,n}$$

and, using factorial notation

$$\left[(n - m)_k + \binom{k}{1}(n - m)_{k-1} mE_m^{-1} + \cdots + \binom{k}{j}(n - m)_{k-j}(m)_j E_m^{-j} + \cdots + (m)_k E_m^{-k}\right] R_{m,n} = (n)_k E_n^{-k} R_{m,n} \tag{17}$$

Symbolically, (17) is

$$[(n - m) + (m)E_m^{-1}]^k R_{m,n} = (n)_k E_n^{-k} R_{m,n}$$

with

$$(n - m)^k \equiv (n - m)_k, \qquad (m)^k \equiv (m)_k$$

Finally, it may be noticed that

$$\Sigma R_{m,n}(x)t^m/m! = (1 + xt)^n \exp t \tag{18}$$

$$\Sigma\Sigma R_{m,n}(x)t^m u^n/m!n! = \exp(t + u + xtu) \tag{19}$$

## 5. CARD MATCHING

As noted before, rectangular polynomials have a particular interest for card matching. Card matching here means the enumeration of the number of matched cards in randomly ordered decks of cards. For two decks, a match consists of like cards in the same position in each deck. For more than two decks, other possibilities are open; a match may consist of like cards in the same position in all decks (as in Problem 3.9), or in at least $k$ of $n$ decks, and so on. For the general case, the specification of the decks (the number of cards of each kind) is left arbitrary.

Considering only two decks, notice first that there is no loss of generality in supposing the total number of cards to be the same in each deck since the smaller may be filled out with new cards different from the old without altering anything, as there are no added matches. Next, since relative rather than absolute position of the cards is important, one deck may be put in standard order and matched against all possible orders of the second. This standard order specifies the positions in which matches can occur and the specification of the second deck specifies the elements which can match in these positions; if there are $s$ kinds of cards to be matched and $n_i$ of the $i$th kind in the first deck, $m_i$ in the second, the array for matches consists of $s$ disjunct rectangles, the $i$th with dimensions $n_i$, $m_i$.

Hence the rook polynomial is

$$R(x) = R_{n_1,m_1}(x)R_{n_2,m_2}(x) \cdots R_{n_s,m_s}(x) \tag{20}$$

The simplest case is for two decks, each of cards numbered 1 to $n$; the enumerating function is $D_n(t)$, corresponding to the rook polynomial $(R_{11})^n$. For the ordinary deck, four suits each with 13 cards, the match may be by number or by suit (the face cards may be given numbers 11, 12, 13). The match by number calls for rook polynomial $[R_{44}(x)]^{13}$, the match by suit for $[R_{13,13}(x)]^4$. Direct calculation for either of these, though straightforward, is formidable and is usually replaced by approximations which will be given later. The procedure can be illustrated in a simpler case, namely:

*Example* 6. (*a*) Consider two decks each with $p$ cards marked $a$, and $q$ cards marked $b$; that is, the specification for each deck is $(pq)$. Then the rook polynomial is

$$R_{pp}R_{qq} \qquad \text{or} \qquad S_pS_q$$

if $S_p = R_{pp}$, the polynomial of a square. For $p = q = 2$

$$S_2(x) = 1 + 4x + 2x^2$$
$$S_2^2(x) = 1 + 8x + 20x^2 + 16x^3 + 4x^4$$

and the hit polynomial is

$$N_{22}(t) = 4! + 8 \cdot 3!\,(t - 1) + 20 \cdot 2(t - 1)^2 + 16(t - 1)^3 + 4(t - 1)^4$$
$$= 4(1 + 4t^2 + t^4)$$

It will be noticed that much of the numerical work is "wasted" in obtaining zero coefficients for $N_{22}(t)$, and direct consideration of the general case ($p, q$ arbitrary) in this way is unattractive.

(*b*) A simpler approach (in this instance) is as follows. Suppose the first deck to be in standard order, and the second deck to have $k$ matches of $a$ cards. Then the second deck must have $p - k$ $b$ cards in $a$ positions, $p - k$ $a$ cards in $b$ positions, and $q - p + k$ matches of $b$ cards, as indicated in the following table:

| | $a$ | $b$ |
|---|---|---|
| $a$ | $k$ | $p - k$ |
| $b$ | $p - k$ | $q - p + k$ |

Column heads are position indicators, row heads card indicators.

The number of matches is the sum of diagonal entries, namely, $q - p + 2k$. The number of ways in which this number of matches can be attained is

$$p!\,q!\binom{p}{k}\binom{q}{p-k}$$

and the enumerating function for number of matches is

$$N_{pq}(t) = p!\,q!\,\Sigma\binom{p}{k}\binom{q}{p-k}t^{q-p+2k}$$

$$= p!\,q!\,\Sigma\binom{p}{k}\binom{q}{k}t^{p+q-2k}$$

The result for $N_{22}(t)$ given above is readily verified as are the following simple instances, in which $N_{pq}(t) = p!\,q!\,A_{pq}(t)$:

$$A_{01} = t$$
$$A_{02} = t^2 \qquad A_{11} = 1 + t^2$$
$$A_{03} = t^3 \qquad A_{12} = 2t + t^3$$
$$A_{04} = t^4 \qquad A_{13} = 3t^2 + t^4 \qquad A_{22} = 1 + 4t^2 + t^4$$

By symmetry $A_{pq}(t) = A_{qp}(t)$; and by the recurrence for binomial coefficients

$$A_{pq}(t) = tA_{p,q-1}(t) + tA_{p-1,q}(t) + (1 - t^2)A_{p-1,q-1}(t)$$

(*c*) Using equation (13) and the result above, it is found that

$$\binom{p+q}{p} S_p(x)S_q(x) = \Sigma\binom{p}{k}\binom{q}{k}R_{p+q,\,p+q-2k}(x)$$

and with the Laguerre polynomial relation mentioned before, namely,

$$R_{n,n+a}(x) = n!\,x^n L_n^a(-x^{-1})$$

this becomes

$$(p+q)!\,L_p(x)L_q(x) = \Sigma\binom{p}{k}\binom{q}{k}(p+q-2k)!\,x^{2k}L_{p+q-2k}^{2k}(x)$$

when, as usual, $L_p(x)$ is written for $L_p^0(x)$.

## 6. CARD MATCHING. APPROXIMATION

As mentioned above, the straightforward enumeration of matches for two decks with a large number of cards is excessively long, and approximate results are desirable. They are achieved by using the fact that the rook polynomial is effectively a generating function for factorial (or binomial) moments of the probability distribution $N_n(t)/N_n(1)$ associated with the hit polynomial, as already noticed in equation (2*b*). Hence, these moments are determined from the rook polynomial, and the approximate distribution is determined by matching as many moments as is necessary.

To find expressions for the moments of card-matching distributions, note first that the rectangular rook polynomial $R_{n,m}(x)$ of equation (10) may be written

$$R_{n,m}(x) = \exp xA, \qquad A^k \equiv A_k = (n)_k(m)_k$$

with $A_k = 0$ for $k$ greater than the lesser of $m$ and $n$. Then the rook polynomial for matching decks of specifications $(n_1, n_2, \cdots, n_s)$, $(m_1, m_2, \cdots, m_s)$, by equation (20), is

$$R(x) = \exp x(a_1 + a_2 + \cdots + a_s), \qquad a_i^k \equiv a_{ik} = (n_i)_k(m_i)_k$$

and, with $n = n_1 + n_2 + \cdots + n_s = m_1 + m_2 + \cdots + m_s$, and $B_k$ the $k$th binomial moment, as in the equation preceding (2b)

$$k!\, r_k = k!\,(n)_k B_k = (a_1 + a_2 + \cdots + a_s)^k, \qquad a_i^k \equiv (n_i)_k(m_i)_k \quad (21)$$

In particular

$$\begin{aligned}
nB_1 &= a_{11} + a_{21} + \cdots + a_{s1} \\
&= n_1m_1 + n_2m_2 + \cdots + n_sm_s \\
2n(n-1)B_2 &= a_{12} + \cdots + a_{s2} + 2(a_{11}a_{21} + \cdots + a_{s-1,1}a_{s1}) \\
&= n_1(n_1-1)m_1(m_1-1) + \cdots + n_s(n_s-1)m_s(m_s-1) \\
&\quad + 2[n_1m_1n_2m_2 + \cdots + n_{s-1}m_{s-1}n_sm_s]
\end{aligned}$$

The expressions of higher moments are increasingly elaborate, and not worth writing out, since, in particular cases, simplifications are made best in (21), as in the following example.

*Example 7.* Suppose both decks of specification $(a^s)$, that is, with $s$ kinds of cards and $a$ of each kind. Then

$$R(x) = [S_a(x)]^s = (\exp bx)^s, \qquad b_k = (a)_k^2$$

and, by the result of Problem 2.22, with $n = as$

$$k!\,(n)_k B_k = Y_k(fb_1, fb_2, \cdots, fb_k), \qquad f^j = (s)_j$$

In particular

$$\begin{aligned}
nB_1 &= sa^2 = na \\
2n(n-1)B_2 &= s(a(a-1))^2 + s(s-1)a^4 \\
&= n(n-1)a^2 - n(a)_2
\end{aligned}$$

from which it appears that $a^k/k!$ is the dominant term in $B_k$. If this were so, then by equation (2.32), namely

$$p_j = \sum_{k=0} (-1)^k \binom{k+j}{k} B_{j+k}$$

with $p_j = N_{nj}/n!$, the probability of $j$ hits, this probability would be $e^{-a}a^j/j!$, the typical term of a Poisson distribution. This suggests approximation by modifying a Poisson distribution through the following expansion for the binomial moments

$$k!\,B_k = A_{k0} + A_{k1}/n + A_{k2}/n(n-1) + \cdots$$

with $A_{ki}$ a function of $a$ as well as of $k$, but independent of $n$. Noting that

$$Y_k(fb_1, \cdots, fb_k) = f_k b_1^k + \binom{k}{2} f_{k-1} b_2 b_1^{k-2} + \left[\binom{k}{3} b_3 b_1 + 3\binom{k}{4} b_2^2\right] f_{k-2} b_1^{k-4} + \cdots$$

it is found first that, with $n(n-a)\cdots(n-(k-1)a) = (n)_{k,a}$,

$$k!(n)_k B_k = a^k(n)_{k,a} + \binom{k}{2} a^{k-1}(a-1)^2(n)_{k-1,a} + \left[\binom{k}{3} a^{k-1}(a-1)^2(a-2)^2 + 3\binom{k}{4} a^{k-2}(a-1)^4\right](n)_{k-2,a} + \cdots$$

and, then, by Problem 2.23,

$$k!\,(n)_k B_k = a^k(n)_k + a^{k-1}(1-a)\binom{k}{2}(n-1)_{k-1} + A_{k2}(n-2)_{k-2} + \cdots$$

with

$$A_{k2} = a^{k-2}\left[a(1-a)\binom{k}{2} + 2(a-1)(a-2)\binom{k}{3} + 3(a-1)^2\binom{k}{4}\right]$$

This is the expansion required, and may be used in equation (2.32), as quoted above, to give

$$p_j = \frac{N_{n,j}}{n!} = \frac{e^{-a}a^j}{j!}\left[1 + \frac{(a-1)[j-(j-a)^2]}{2a(n-1)} + \frac{(a-1)f_2(a,j)}{24a^2n(n-1)} + \cdots\right]$$

where

$$f_2(a,j) = 8(a-2)[(j)_3 - 3a(j)_2 + 3a^2j - a^3] + 3(a-1)[(j)_4 - 4a(j)_3 + 6a^2(j)_2 - 4a^3j + a^4]$$

This is an improvement on a formula given by Cattaneo, **2**, which is in powers of $n^{-1}$.

## 7. COMPLEMENTS

The complement of a chessboard in any board which includes it is the board formed of all cells not in the given board. Thus the two boards

```
×          × ×
  ×        ×   ×
    ×      × ×
```

are complements in a square of side three.

The polynomials of boards which are complements in a rectangle are related as follows. Take $Q(x)$ and $R(x)$ for the two polynomials, complementary in an $m$ by $n$ rectangle; then

$$\begin{aligned} Q(x) &= \Sigma r_k(-x)^k R_{m-k,n-k}(x) \\ &= R(-xf)R_{m,n}(x), \qquad f^k R_{m,n}(x) = R_{m-k,n-k} \end{aligned} \tag{22}$$

Equation (22) may be obtained directly from the principle of inclusion and exclusion, Theorem 3.1, if the argument is adapted to polynomials rather than to their numerical coefficients as stated. Thus $Q(x)$ corresponds to $N(a'b'c' \cdots)$, $N = N(1)$ to the polynomial $R_{nm}(x)$ of the whole board, $N(a)$ to the polynomial with a rook on one cell of the board with polynomial $R(x)$, which is $xR_{n-1,m-1}(x)$, and so on. As there are $r_k$ ways of putting $k$ rooks on the board with polynomial $R(x)$, and as the polynomial for a rectangle with $k$ rooks in given (compatible) cells is $x^k R_{n-k,m-k}(x)$, the general term of (22) is clear.

*Example 8.* For the boards shown above, the rook polynomial of the first is $(1 + x)^3$. Hence that of the second is, with $S_n(x)$ the polynomial of a square of side $n$,

$$\begin{aligned} Q(x) &= S_3(x) - 3xS_2(x) + 3x^2S_1(x) - x^3S_0(x) \\ &= 1 + 9x + 18x^2 + 6x^3 - 3x(1 + 4x + 2x^2) + 3x^2(1 + x) - x^3 \\ &= 1 + 6x + 9x^2 + 2x^3 \end{aligned}$$

This board is the case $n = 3$ of the reduced "problème des ménages." Conversely

$$R(x) = (1 + x)^3 = S_3 - 6xS_2 + 9x^2S_1 - 2x^3$$

It is sometimes convenient to find a complement in steps, or by recurrence. Thus, if $R(x)$ is given in the form

$$R(x) = R_0(x) + xR_1(x) + \cdots + x^jR_j(x) + \cdots$$

then its complement in an $m$ by $n$ rectangle is

$$Q(x) = Q_0(x) - xQ_1(x) + \cdots + (-x)^jQ_j(x) + \cdots \tag{23}$$

where $Q_j(x)$ is the complement of $R_j(x)$ in the rectangle $m - j$, $n - j$.

If $R(x)$ and $Q(x)$ are complementary polynomials on a square of side $n$, and the corresponding hit polynomials for $n$ elements are

$$N_n(t) = \Sigma N_{n,j}t^j, \qquad M_n(t) = \Sigma M_{n,j}t^j$$

then $N_{n,j} = M_{n,n-j}$, because, when $j$ elements are in forbidden positions on one board, $n - j$ are in forbidden positions on the complementary board. The corresponding polynomial relations are

$$\begin{aligned} N_n(t) &= t^n M_n(t^{-1}) \\ M_n(t) &= t^n N_n(t^{-1}) \end{aligned} \tag{24}$$

For the rook polynomials of Example 8, the hit polynomials are $N_3(t) = D_3(t) = 2 + 3t + t^3$ and $M_3(t) = 1 + 3t^2 + 2t^3$.

Using equation (3), the first of (24) becomes

$$R[(t - 1)E^{-1}]n! = t^n Q[(t^{-1} - 1)E^{-1}]n!$$

or, with $x = t - 1$

$$R(xE^{-1})n! = (1 + x)^n Q[-xE^{-1}(1 + x)^{-1}]n! \tag{25}$$

Equation (25) may be restated as:

**Theorem 2.** *If $R(x)$ and $Q(x)$ are rook polynomials of two boards complementary in a square of side $n$ and $E^{-k}n! = (n - k)!$, then*

$$R(xE^{-1})n! = (1 + x)^n Q[-xE^{-1}(1 + x)^{-1}]n!$$

*or*

$$\Sigma r_k x^k (n - k)! = \Sigma q_k (-x)^k (1 + x)^{n-k} (n - k)!$$

## 8. EQUIVALENCE

As already noticed, boards that may be made alike by permutation of their rows or columns or by interchanging rows with columns have the same polynomials, hence are equivalent, a relation for which the usual symbol is $\sim$. Thus

```
  × ×   ~   ×
            ×

× × ×   ~   × ×     ~   × ×
  × ×       × × ×       × × ×
    ×       ×             ×
```

There are other less obvious cases of equivalence; for example,

(*a*)

```
  × ×   ~   × × ×
  × ×           ×
```

(*b*)

```
× × ×   ~   × × ×   ~   × × × ×
  × ×           ×               ×
                ×
```

(*c*)

```
× × ×   ~   × ×
×           × ×
  ×           ×
    ×           ×
```

(*d*)

```
  × ×   ~   ×
×   ×       ×
× ×         × ×
              × ×
```

These boards are not geometrically congruent and yet have the same polynomials. The polynomials are: (*a*) $1 + 4x + 2x^2$; (*b*) $1 + 5x + 4x^2$; (*c*) $1 + 6x + 9x^2 + 4x^3$; (*d*) $1 + 6x + 9x^2 + 2x^3$. Note that (*b*) and (*c*) are factorable and that both factors are rook polynomials; in contrast, (*d*) may be factored into $(1 + 4x + x^2)(1 + 2x)$ but $(1 + 4x + x^2)$ is not a rook polynomial.

When are two polynomials equivalent? A complete answer has not been found, but some information is given in the following theorems. The first is due to Irving Kaplansky (private communication) and requires a slight preliminary. Consider the $m$ by $n + p$ rectangle

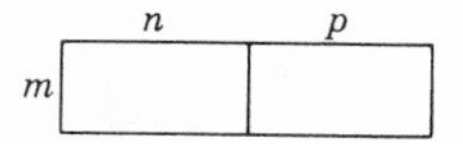


and suppose that the $m$ by $n$ space is a completely filled board, board $C$, with rook polynomial $R_{m,n}(x)$, and that two equivalent boards $A$ and $B$ may each be fitted into the $m$ by $p$ space. Then:

**Theorem 3.** *If $A$ and $B$ are equivalent boards, that is, boards with the same rook polynomials, so are $A + C$ and $B + C$.*

$A + C$ is the board consisting of the cells of $A$ and the $(mn)$ cells of $C$. Equivalence (*a*) and the first of (*b*) are examples of this theorem.

Proof is as follows. Write

$$R(x, A) = R(x, B) = \Sigma r_k x^k$$

$$R(x, A + C) = \Sigma R_k x^k$$

Then

$$R_k = \sum_{j=0}^{k} r_j \binom{m-j}{k-j}\binom{n}{k-j}(k-j)!$$

since with $j$ rooks on board $A$, no rooks may be in the same rows on board $C$, and

$$\binom{m-j}{k-j}\binom{n}{k-j}(k-j)!$$

is the number of ways of putting $k - j$ rooks on an $m - j$ by $n$ rectangle. Clearly

$$R(x, B + C) = \Sigma R_k x^k = R(x, A + C)$$

as stated in the theorem.

The second theorem includes the first (Theorem 3) and is as follows:

**Theorem 4.** *If two polynomials are equivalent, so are their complements in any rectangle.*

Proof is immediate from equation (22). The proof that it includes Theorem 3 is as follows. Take the completely filled $m$ by $p$ rectangle as board $D$, and write the complement of $A$ on this board as $D - A$ and of $B$ as $D - B$. By Theorem 4

$$D - A \sim D - B$$

the equivalence referring to either the boards or their rook polynomials. The complement of $D - A$ on the $m$ by $n + p$ rectangle $C + D$ is $A + C$, the complement of $D - B$ on the same rectangle is $A + B$ and, by Theorem 4 again, $A + B \sim A + C$, as stated in Theorem 3.

The third theorem is somewhat different, and almost self-evident.

**Theorem 5.** *Two polynomials are equivalent if both inclusion and exclusion polynomials on single-cell expansion are the same.*

Case ($d$) above is an instance of an equivalence provable by Theorem 5 but not by Theorem 4. The single-cell expansions are

$$\begin{bmatrix} & \times & \times \\ \times & & \times \\ \times & \times & \end{bmatrix} = \begin{bmatrix} & & \times \\ \times & & \times \\ \times & \times & \end{bmatrix} + x \begin{bmatrix} \times & \times \\ \times & \end{bmatrix}$$

$$\begin{bmatrix} \times & & \\ \times & & \\ \times & \times & \\ & \times & \times \end{bmatrix} = \begin{bmatrix} \times & & \\ \times & \times & \\ & \times & \times \end{bmatrix} + x \begin{bmatrix} \times & \\ \times & \times \end{bmatrix}$$

The difficulty in applying Theorem 5 is to find suitable cells for expansion. In this respect, it is handy to have the

*Definition. Two cells of a given board are equivalent for expansion if either their inclusion or exclusion boards are equivalent.*

By equation (7), since the polynomial of a single board is being expanded, if the inclusion boards are equivalent, so are the exclusion boards (and vice versa).

Table 1 shows all distinct polynomials and their boards for connected boards with at most six cells. Notice that $1 + nx$ is the only binomial for any $n$, and that the "largest" polynomial, $L_n(x)$, that is, the one with the greatest sum of coefficients, is determined by $L_0(x) = 1$, $L_1(x) = 1 + x$, and the recurrence

$$L_n(x) = L_{n-1}(x) + xL_{n-2}(x) \tag{26}$$

$L_n(x)$ has a board which is a staircase (the board for the reduced ménage problem when the table is straight rather than circular, as will appear).

The proof that this is so follows from expansion with respect to a single cell and mathematical induction. First, $L_1(x) = 1 + x$ is the largest (and only) polynomial for a single cell. Next, any polynomial for $n$ connected cells by single cell expansion must be of the form

$$R_n(x) = P_{n-1}(x) + xQ_{n-k}(x), \qquad k > 1$$

with the subscripts indicating number of cells, since the inclusion of a given cell must exclude at least one other cell in a connected board. Hence, if $L_j(x)$ is the largest polynomial for any $j$ less than $n$, $L_n(x)$ is the largest polynomial for $n$ [choosing $L_{n-1}(x)$ for $P_{n-1}(x)$ and $L_{n-2}(x)$ for $Q_{n-k}(x)$].

Notice that (26) and the initial conditions imply that

$$L_n(x) = 1 + nx + \binom{n-1}{2}x^2 + \cdots + \binom{n-k+1}{k}x^k + \cdots \binom{n-m+1}{m}x^m \tag{27}$$

with $m = [(n + 1)/2]$.

## REFERENCES

1. I. L. Batten, On the problem of multiple matching, *Annals of Math. Statist.*, vol. 13 (1942), pp. 294–305.
2. P. Cattaneo, Sul problema delle concordanze, *Inst. Veneto Sci. Lett. Arti. Parte II, Cl. Sci. Mat. Nat.*, vol. 101 (1942), pp. 89–104.
3. M. Frechet, A note on the problème des rencontres, *Amer. Math. Monthly*, vol. 46 (1939), p. 501.
4. ———, *Les probabilités associées a un system d'évènements compatibles et dépendants*, part II, Paris, 1943.
5. I. Kaplansky, On a generalization of the "problème des rencontres", *Amer. Math. Monthly*, vol. 46 (1939), pp. 159–161.
6. ———, Symbolic solution of certain problems in permutations, *Bull. Amer. Math. Soc.*, vol. 50 (1944), pp. 906–914.
7. ——— and J. Riordan, Multiple matching and runs by the symbolic method, *Annals of Math. Statist.*, vol. 16 (1945), pp. 272–277.
8. ———, The problem of the rooks and its applications, *Duke Math. Journal*, vol. 13 (1946), pp. 259–268.
9. S. Kullback, Note on a matching problem, *Annals of Math. Statist.*, vol. 10 (1939), pp. 77–80.
10. L. Lindelöf, Opersigt af Finska Vetenskaps, *Societetens Förhandlingar*, vol. 42 (1899–1900).
11. E. Lucas, *Théorie des nombres*, Paris, 1891, pp. 491–495.
12. N. S. Mendelsohn, Symbolic solution of card matching problems, *Bull. Amer. Math. Soc.*, vol. 52 (1946), pp. 918–924.
13. E. G. Olds, A moment generating function which is useful in solving certain matching problems, *Bull. Amer. Math. Soc.*, vol. 44 (1938), pp. 407–413.
14. S. S. Wilks, *Mathematical Statistics*, Princeton, 1943, chapter X.
15. K. Yamamoto, Structure polynomial of Latin rectangles and its application to a combinatorial problem, *Mem. Fac. Sci. Kyusyu Univ.*, ser. A, vol. 10 (1956), pp. 1–13.

## PROBLEMS

1. (*a*) Show that the rook polynomial of a square of side $n$, $S_n(x)$, satisfies the recurrences

$$S_{n+1}(x) = [1 + (2n + 1)x]S_n(x) - n^2x^2S_{n-1}(x)$$
$$S_n'(x) = (d/dx)S_n(x)$$
$$= n^2S_{n-1}(x), \qquad n > 0$$

Verify the evaluations:

$$S_1 = 1 + x \qquad S_3 = 1 + 9x + 18x^2 + 6x^3$$
$$S_2 = 1 + 4x + 2x^2 \qquad S_4 = 1 + 16x + 72x^2 + 96x^3 + 24x^4$$

(*b*) Using these recurrences, find the exponential generating function relations

$$(1 - xt)^2\, D \exp tS(x) = (1 + x - x^2t) \exp tS(x), \qquad D = d/dt$$
$$\exp tS'(x) = (t + t^2D) \exp tS(x)$$
$$(1 - xt) \exp tS(x) = \exp [t(1 - xt)^{-1}]$$

2. The complement of an $m$ by $n$ rectangle in an $m$ by $(n + p)$ rectangle is an $m$ by $p$ rectangle. Hence, by equation (21)

$$R_{m,p}(x) = \Sigma(-x)^k \binom{m}{k}\binom{n}{k} k!\, R_{m-k,n+p-k}(x)$$

Verify by iterating the rectangular polynomial recurrence, (9*a*),

$$R_{m,n-1}(x) = R_{m,n}(x) - mxR_{m-1,n-1}(x)$$

3. A permanent of order $n$ may be defined by $\Sigma a_{1i_1}a_{2i_2}\cdots a_{ni_n}$ with $i_1, i_2, \cdots, i_n$ a permutation of numbers 1 to $n$; it is like a determinant without sign changes and is represented by

$$\overset{+}{\phantom{x}}\begin{vmatrix} a_{11} & \cdots & a_{1n} \\ a_{21} & \cdots & a_{2n} \\ a_{n1} & \cdots & a_{nn} \end{vmatrix}\overset{+}{\phantom{x}}$$

Show that $D_n(t)$, the hit polynomial for rencontres, is the permanent of order $n$ with $t$ on the main diagonal and unity elsewhere; for example,

$$D_2(t) = \overset{+}{\phantom{x}}\begin{vmatrix} t & 1 \\ 1 & t \end{vmatrix}\overset{+}{\phantom{x}} = 1 + t^2$$

$$D_3(t) = \overset{+}{\phantom{x}}\begin{vmatrix} t & 1 & 1 \\ 1 & t & 1 \\ 1 & 1 & t \end{vmatrix}\overset{+}{\phantom{x}} = tD_2(t) + 2(1 + t) = 2 + 3t + t^3$$

Note that the $t$ positions are the rencontres board.

*Hint*: Derive the recurrences

$$D_n(t) = tD_{n-1}(t) + (n - 1)f_{n-1}(t)$$
$$f_{n-1}(t) = D_{n-2}(t) + (n - 2)f_{n-2}(t)$$

4. Show that the cells on the hypotenuse of a triangular board are equivalent, and that the same is true for cells on any line parallel to the hypotenuse.

5. Show that the coefficients $r_k$ with odd index of the rook polynomial $R(x)$ which is self-complementary in an $m$ by $n$ rectangle must satisfy relations like

$$2r_1 = mn$$
$$24r_3 = 12(m-2)(n-2)r_2 - (m)_3(n)_3$$

but that the even index coefficients are undetermined. For example, for a 2 by 2 rectangle, both boards

$$\begin{matrix} \times & \times \end{matrix} \quad \text{and} \quad \begin{matrix} \times & \\ & \times \end{matrix}$$

are self-complementary but with different polynomials ($1 + 2x$ and $1 + 2x + x^2$). For a 2 by 4 rectangle, the two boards

$$\begin{matrix} \times & \times \\ \times & \times \end{matrix} \qquad \begin{matrix} \times & \times & \times \\ \times & & \end{matrix}$$

have the same polynomial ($1 + 4x + 2x^2$) but the board

$$\begin{matrix} \times & \times & & \\ & & \times & \times \end{matrix}$$

also self-complementary, has polynomial $1 + 4x + 4x^2$.

6. A simple staircase is taken in the form

$$\begin{matrix} \times & \times & & & & & \\ & \times & \times & & & & \\ & & \times & \times & & & \\ & & & & \cdots & & \end{matrix}$$

and, hence, has polynomial $(1 + 2x)^m$ for $m$ "steps".

(*a*) Verify the following hit polynomials ($n = 2m$):

$$A_2(t) = 2t$$
$$A_4(t) = 8(1 + t + t^2)$$
$$A_6(t) = 48(5 + 6t + 3t^2 + t^3)$$

(*b*) Write $A_{2m}(t) = 2^m m!\, a_m(t)$; show that with a prime denoting a derivative,

$$\begin{aligned} a_m(t) &= (2m-1)a_{m-1}(t) + (1-t)^2 a_{m-2}(t) \\ &= (2m-2+t)a_{m-1}(t) + (1-t)a'_{m-1}(t) \\ &= a'_m(t) - (1-t)a_{m-1}(t) \end{aligned}$$

*Hint*: Use the relation

$$\binom{m}{k}(2m-k)! = (2m-1)2m\binom{m-1}{k}(2m-2-k)! + m(m-1)\binom{m-2}{k-2}(2m-2-k)!$$

(*c*) Show that $a_m(t)$ is of degree $m$ and, with

$$a_m(t) = \Sigma a_{m,k} t^k$$

$$a_{m,m} = 1, \qquad a_{m,m-1} = \binom{m}{2}, \qquad a_{m,m-2} = \binom{m}{2} + 3\binom{m+1}{4}$$

(*d*) From (*b*) verify that $a_m(1) = (2m)!/m!\,2^m = a'_m(1)$, with a prime denoting a derivative, and that

$$a''_{m-1}(1) = a'_m(1) - a_{m-1}(1)$$

Hence, the first two factorial moments are

$$m_1 = 1, \qquad (m)_2 = (2m-2)/(2m-1)$$

7. For matching two decks of cards each of specification $2^m$ $(n = 2m)$, that is, $m$ kinds and two of each kind, the rook polynomial is $(1 + 4x + 2x^2)^m$.

(*a*) With $A_{2m}(t) = 2^m a_m(t)$, the corresponding hit polynomial, verify that

$$a_1(t) = t^2$$
$$a_2(t) = 1 + 4t^2 + t^4$$
$$a_3(t) = 10 + 24t + 27t^2 + 16t^3 + 12t^4 + t^6$$

(*b*) Show that, with primes denoting derivatives,

$$a'_m(t) = 2m(2m - 2 + t)a_{m-1}(t) + 2m(1 - t)\,a'_{m-1}(t)$$
$$2a_m(t) = [(2m-2)(2m-3) + 8(m-1)t + 2t^2]a_{m-1}(t)$$
$$+ (4m - 6 + 4t)(1 - t)\,a'_{m-1}(t) + (1 - t)^2\,a''_{m-1}(t)$$

*Hint*: follow the procedure of Example 5(*a*).

8. *Continuation.* (*a*) Show that, with $Eu_n = u_{n+1}$, and $D_n$ the displacement number

$$e_n = (E - 2)^n 0! = (D - 1)^n$$
$$\exp ue = e^{-2u}(1 - u)^{-1}, \qquad e^n \equiv e_n$$
$$\exp u(e + 2t) = (1 - u)^{-1} \exp 2u(t - 1)$$
$$e_n(2t) = (e + 2t)^n, \qquad e^k \equiv e_k$$
$$= ne_{n-1}(2t) + (2t - 2)^n$$
$$= n(n-1)e_{n-2}(2t) + (n - 2 + 2t)(2t - 2)^{n-1}$$

(*b*) Using the relation

$$A_{2m}(t) = [E^2 + 4E(t - 1) + 2(t - 1)^2]^m 0!$$
$$= [e(2t) - 2(t - 1)^2]^m, \qquad e^k(2t) \equiv e_k(2t)$$

show that

$$A_{2m}(t) = 2m(2m-1)A_{2m-2}(t) + 8m(m-1)(1-t)^2 A_{2m-4}(t)$$
$$+ m2^{m+1}(t-1)^{2m-1} + 2^m(t-1)^{2m}$$
$$a_m(t) = m(2m-1)a_{m-1}(t) + 2m(m-1)(1-t)^2 a_{m-2}(t)$$
$$+ (2m - 1 + t)(t - 1)^{2m-1}$$

(*c*) Verify that $a_m(1) = (2m)!\,2^{-m}$, and show that, with a prime indicating a derivative,

$$a'_m(1) = 2a_m(1)$$
$$a''_m(1) = 2ma_{m-1}(1) + 2m(2m-2)a'_{m-1}(1)$$

so that

$$m_1 = 2$$
$$(m)_2 = m_2 - m_1 = (8m-6)/(2m-1)$$

(*d*) Show that the binomial moments $B_k(m)$ satisfy

$$(2m-1)(k+1)B_{k+1}(m) = 2(2m-1-k)B_k(m-1) + 2B_{k-1}(m-1)$$

so that, for large $m$,

$$B_k(m) \sim 2^k/k!$$

*Hint*: $B(t, m) = \Sigma B_k(m)t^k = a_m(1+t)/a_m(1)$; use the first relation in Problem 7(*b*).

9. (*a*) Show that for $n = 2m$, the number of terms in the expansion of a determinant of order $n$ which contain no elements from either of the two main diagonals is $A_{2m}(0)$, with $A_{2m}(t)$ as in Problem 8(*b*), and, hence, that

$$A_{2m}(0) = 2m(2m-1)A_{2m-2}(0) + 8m(m-1)A_{2m-4}(0) + (1-2m)2^m$$
$$a_m(0) = m(2m-1)a_{m-1}(0) + 2m(m-1)a_{m-2}(0) + (1-2m)$$

The first few values are

| $m$ | 1 | 2 | 3 | 4 | 5 | 6 |
|---|---|---|---|---|---|---|
| $A_{2m}(0)$ | 0 | 4 | 80 | 4752 | 440192 | 59245120 |
| $a_m(0)$ | 0 | 1 | 10 | 297 | 13756 | 925705 |

(*b*) For $n = 2m+1$, the rook polynomial for the problem of 9(*a*) is $(1+4x+2x^2)^m(1+x)$. Show that the corresponding polynomial $A_{2m+1}(t)$ satisfies, with a prime indicating a derivative,

$$4(m+1)A_{2m+1}(t) = A'_{2m+2}(t)$$
$$A_{2m+1}(t) - 4m(1-t)A_{2m-1}(t) = (2m+t)A_{2m}(t)$$

Hence

$$A_{2m+1}(0) = 2m[2A_{2m-1}(0) + A_{2m}(0)]$$

Verify the following:

$$A_1(t) = t \qquad A_1(0) = 0$$
$$A_3(t) = 4t + 2t^3 \qquad A_3(0) = 0$$
$$A_5(t) = 4(4 + 9t + 8t^2 + 8t^3 + t^5) \qquad A_5(0) = 16$$
$$A_7(0) = 528$$

10. The rook polynomials, $Q_n(x)$, for the board complementary to the rencontres board in a square of side $n$ are defined by

$$Q_n(x) = \sum_0^n \binom{n}{k} (-x)^k S_{n-k}(x)$$
$$= (S-x)^n, \qquad S^n = S_n(x) = \text{square polynomial}$$

(*a*) Show that

$$\begin{aligned}\exp tQ(x) &= \exp(-xt)\exp tS(x)\\(1-xt)^2Q(x)\exp tQ(x) &= (1+x^2t-x^3t^2)\exp tQ(x)\\\exp tQ'(x) &= t^2Q(x)\exp tQ(x)+t^2x\exp tQ(x)\end{aligned}$$

and, hence, that

$$\begin{aligned}Q_{n+1}(x) &= (1+2nx)Q_n(x)-n(n-2)x^2Q_{n-1}(x)-n(n-1)x^3Q_{n-2}(x)\\Q_n'(x) &= n(n-1)[Q_{n-1}(x)+xQ_{n-2}(x)]\end{aligned}$$

Verify these recurrences using the initial values

$$Q_0 = Q_1 = 1, \qquad Q_2 = (1+x)^2, \qquad Q_3 = 1+6x+9x^2+2x^3$$

(*b*) Write $Q_n(x) = \Sigma Q_{n,k}x^k$; show that

$$\begin{aligned}Q_{n,k} &= \binom{n}{k}\Delta^k(n-k)!/(n-k)!\\&= \binom{n}{k}D^k(D+1)^{n-k}/(n-k)!, \qquad D^n \equiv D_n = \Delta^n 0!\end{aligned}$$

Verify the instances

$$Q_{n0} = 1, \qquad Q_{n1} = n(n-1), \qquad Q_{n2} = \binom{n}{2}(n^2-3n+3), \qquad Q_{nn} = D_n$$

and the relation

$$kQ_{n,k} = n(n-1)(Q_{n-1,k-1}+Q_{n-2,k-2})$$

(*c*) Write $Q_{n,k} = \binom{n}{k}d_{n,n-k}$ so that

$$d_{n,k} = \Delta^{n-k}k!/k!$$

Show that

$$\begin{aligned}kd_{n,k} &= d_{n,k-1}+d_{n-1,k-1}\\d_{n,k} &= (n-k)\,d_{n-1,k}+d_{n-1,k-1}\\&= (n-1)\,d_{n-1,k}+(n-k-1)d_{n-2,k}\end{aligned}$$

Verify the table:

| $n\backslash k$ | 0 | 1 | 2 | 3 | 4 | 5 | 6 | 7 |
|---|---|---|---|---|---|---|---|---|
| 0 | 1 | | | | | | | |
| 1 | 0 | 1 | | | | | | |
| 2 | 1 | 1 | 1 | | | | | |
| 3 | 2 | 3 | 2 | 1 | | | | |
| 4 | 9 | 11 | 7 | 3 | 1 | | | |
| 5 | 44 | 53 | 32 | 13 | 4 | 1 | | |
| 6 | 265 | 309 | 181 | 71 | 21 | 5 | 1 | |
| 7 | 1854 | 2119 | 1214 | 465 | 134 | 31 | 6 | 1 |

11. The hit polynomial corresponding to $Q_n(x)$ is

$$A_n(t) = \Sigma Q_{n,k}(n-k)!\,(t-1)^k$$

(*a*) Show that

$$\begin{aligned}A_n(t) &= \Sigma\binom{n}{k}D_kt^k\\&= (1+Dt)^n, \qquad D^k \equiv D_k\end{aligned}$$

Verify the first few values

$$A_0 = A_1 = 1, \qquad A_2 = 1 + t^2, \qquad A_3 = 1 + 3t^2 + 2t^3$$

(*b*) With $D_n(t)$ the displacement number polynomial, show that

$$A_n(t) = t^n D_n(t^{-1})$$

and hence that

$$\begin{aligned}(1 - ut) \exp uA(t) &= \exp u(1 - t)\\ A_n(t) &= ntA_{n-1}(t) + (1 - t)^n\\ n!\, t^n &= [A(t) + (t - 1)]^n\\ &= \Sigma \binom{n}{k} (t - 1)^k A_{n-k}(t)\end{aligned}$$

(*c*) From the relation

$$n!\, Q_n(x) = \Sigma \binom{n}{k} D_k R_{n,k}(x)$$

verify the expressions for $Q_{n,k}$ in Problem 10(*b*).

12. The rook polynomials $Q_{n,m}(x)$ for the board complementary to the rencontres board in an $n$ by $m$ rectangle, $n \geq m$, are defined by

$$Q_{n,m}(x) = \Sigma \binom{m}{k} (-x)^k R_{n-k,m-k}(x)$$

and in the usual way

$$Q_{n,m}(x) = \Sigma Q_k(n, m) x^k$$

(*a*) Show that, in the notation of Problem 10,

$$Q_k(n, m) = \binom{m}{k} d_{n,n-k}$$

Verify the instances

$$\begin{aligned}Q_{0,0} = Q_{n,0} = 1, \qquad Q_{n,1}(x) &= 1 + (n - 1)x\\ Q_{n,2}(x) = 1 + 2(n - 1)x &+ (n^2 - 3n + 3)x^2\end{aligned}$$

(*b*) Derive the recurrence

$$Q_{n,m}(x) = Q_{n-1,m}(x) + mxQ_{n-1,m-1}(x), \qquad m < n$$

(*c*) The corresponding hit polynomial $A_{n,m}(t)$ is defined by

$$A_{n,m}(t) = \Sigma Q_k(n, m)(n - k)!\, (t - 1)^k$$

Show that

$$A_{n,m}(t) = (D + 1)^{n-m} (1 + tD)^m, \qquad D^k \equiv D_k = \Delta^k 0!$$

Verify the instances

$$\begin{aligned}A_{n1}(t) &= (n - 1)!\, [1 + (n - 1)t]\\ A_{n2}(t) &= (n - 2)!\, [1 + 2(n - 2)t + (n^2 - 3n + 3)t^2]\end{aligned}$$

(*d*) With $A_{nn}(t) = A_n(t)$ of Problem 11, show that

$$\begin{aligned}A_{n,n-1}(t) &= nA_{n-1}(t) - (n - 1)(1 - t)A_{n-2}(t)\\ A_{n+1,n}(t) &= ntA_{n,n-1}(t) + A_n(t)\\ A_{n+2,n}(t) &= ntA_{n+1\; n-1}(t) + 2A_{n+1,n}(t)\end{aligned}$$

(*e*) Write $A_{n+m,n}(t) = A_n^{(m)}(t)$; show that

$$(1 - ut) \exp uA^{(m)}(t) = m \exp uA^{(m-1)}(t)$$
$$(1 - ut)^{m+1} \exp uA^{(m)}(t) = m! \exp [u(1 - t)]$$

and, hence, that

$$A_n^{(m)}(t) = ntA_{n-1}^{(m)}(t) + mA_n^{(m-1)}(t)$$
$$A_{n+1}^{(m)}(t) = [1 + (n + m)t]A_n^{(m)}(t) + n(t^2 - t)A_{n-1}^{(m)}(t)$$

13. By application of Theorem 3, show the following equivalences:

$$\begin{matrix} \times\times\times \\ \times \end{matrix} \sim \begin{matrix} \times\times \\ \times\times \end{matrix}$$

$$\begin{matrix} \times\times\times\times\times \\ \times\times\times \\ \times \end{matrix} \sim \begin{matrix} \times\times\times \\ \times\times\times \\ \times\times\times \end{matrix}$$

and so on. Note that, in the general case, this implies

$$1 + 3 + \cdots + (2n - 1) = n^2$$

14. For the problem of incomplete rencontres (Example 2), write

$$D_n(t, m) = \Sigma D_k(n, m)t^k$$

so that

$$D_k(n, m) = \binom{m}{k} \Delta^{m-k}(n - m)!$$

$$n! (1 + x)^m = \Sigma D_k(n, m)R_{n,k}(x)$$

(*a*) Show that for $m < n$, $n! (1 + x)^m = n[(n - 1)! (1 + x)^m]$ implies

$$D_k(n, m) = (n - k)D_k(n - 1, m) + (k + 1)D_{k+1}(n - 1, m)$$

and hence, with a prime for a derivative,

$$D_n(t, m) = nD_{n-1}(t, m) + (1 - t)D'_{n-1}(t, m)$$

(*b*) Show that

$$D'_n(t, m) = mD_{n-1}(t, m - 1)$$

so that

$$D_n(t, m) = nD_{n-1}(t, m) + m(1 - t)D_{n-2}(t, m - 1), \qquad m \leq n$$

(*c*) Derive the recurrences

$$\begin{aligned} \Delta^{m-k}(n - m)! &= \Delta^{m-k+1}(n - m - 1)! + \Delta^{m-k}(n - m - 1)! \\ &= (n - m)\, \Delta^{m-k}(n - m - 1)! + (m - k)\, \Delta^{m-k-1}(n - m)! \\ &= (n - k)\, \Delta^{m-k}(n - m - 1)! + (m - k)\, \Delta^{m-k-1}(n - m - 1)! \end{aligned}$$

From the second derive the formulas:

$$\begin{aligned} D_n(t, m) &= (n - m)D_{n-1}(t, m) + D'_n(t, m) \\ &= (n - m)D_{n-1}(t, m) + mD_{n-1}(t, m - 1) \end{aligned}$$

*Hint*: for the second recurrence, use the expression

$$\Delta^{m-k}(n - m)! = \Sigma \binom{n - m}{j} D_{n-k-j}$$

and the recurrence

$$D_n = nD_{n-1} + (-1)^n$$

15. The card-matching polynomial of Example 6, for $p = q$, may be written as

$$A_{pp}(t) = \Sigma \binom{p}{k}^2 t^{2k} = a_p(u), \qquad u = t^2$$

Using the relations

$$k^2 \binom{p}{k}^2 = p^2 \binom{p-1}{k-1}^2$$

$$(p-k)^2 \binom{p}{k}^2 = p^2 \binom{p-1}{k}^2$$

$$k \binom{p}{k}^2 = k \binom{p-1}{k}^2 + (2p-k) \binom{p-1}{k-1}^2$$

Show that, with primes denoting derivatives,

$$a_p'(u) + ua_p''(u) = p^2 a_{p-1}(u)$$

$$p^2 a_p(u) - (2p-1)ua_p'(u) + u^2 a_p''(u) = p^2 a_{p-1}(u)$$

$$a_p'(u) = (2p-1)a_{p-1}(u) + (1-u)a_{p-1}'(u)$$

and, hence, that

$$pa_p(u) = (2p-1)(1+u)a_{p-1}(u) - (p-1)(1-u)^2 a_{p-2}(u)$$

Verify the last with the initial values:

$$a_0(u) = 1, \qquad a_1(u) = 1 + u, \qquad a_2(u) = 1 + 4u + u^2$$

Comparing these and the recurrence relation for Legendre polynomials, namely,

$$nP_n(u) = (2n-1)uP_{n-1}(u) - (n-1)P_{n-2}(u)$$

show that

$$a_n(u) = (1-u)^n P_n[(1+u)(1-u)^{-1}]$$

16. For card matching with two decks of specification $(p_1 q_1)$ and $(p_2 q_2)$, show that the hit polynomial is

$$A(t) = \Sigma \binom{p_2}{k} \binom{q_2}{p_1 - k} t^{q_2 - p_2 + 2k}$$

17. *Complement to Example 7.* Write the square polynomial of side $a$ as

$$S_a(x) = x^a a! \exp cx^{-1}, \qquad c^k \equiv c_k$$

and show that $c_k = \binom{a}{k}$. Then, if

$$[S_a(x)]^s = x^{as} a!^s \exp c(s)\, x^{-1}, \qquad c^n(s) \equiv c_n(s)$$

the hit polynomial $N(t, a^s)$ is given by

$$N(t, a^s) = a!^s \sum_{k=0} c_k(s)(t-1)^{as-k} = a!^s \Sigma A_k t^k$$

From $c_0(s) = 1$, $c_1(s) = as$, $c_2(s) = s\binom{a}{2} + s(s-1)a^2$, and succeeding values, show that

$$A_{as} = 1$$
$$A_{as-1} = 0$$
$$A_{as-2} = a^2\binom{s}{2}$$
$$A_{as-3} = 2a^3\binom{s}{3}$$
$$A_{as-4} = \binom{a}{2}^2\binom{s}{2} + 6a^2\binom{a}{2}\binom{s}{3} + 9a^4\binom{s}{4}$$
$$A_{as-5} = 12a\binom{a}{2}^2\binom{s}{3} + 48a^3\binom{a}{2}\binom{s}{4} + 44a^5\binom{s}{5}$$
$$A_{as-6} = \left[2a^2\binom{a}{4} + a\binom{a}{3}\right]\binom{s}{2}\Big/3 + a(11a-17)\binom{a}{2}^2\binom{s}{3}$$
$$+ \left[190a^2\binom{a}{2}^2 - 8a^3\binom{a}{2}\right]\binom{s}{4} + 420a^4\binom{a}{2}\binom{s}{5} + 265a^6\binom{s}{6}$$

18. For two decks each of specification $pq \cdots w$, show that the reduced hit polynomial $A_{pq\cdots w}(t)$ $[A_{pq\cdots w}(1) = (p+q+\cdots w)!/p!\,q!\cdots w!]$ satisfies ($n = p + q + \cdots + w$, $D = d/dt$)

$$pA_{pq\cdots w}(t) = [n + 1 - 2p + t(2p-1)]A_{p-1,q,\cdots,w}(t)$$
$$+ (1-t)\,DA_{p-1,q,\cdots,w}(t)$$
$$- (p-1)(1-t)^2A_{p-2,q,\cdots,w}(t)$$
$$DA_{pq\cdots w}(t) = pA_{p-1,q,\cdots,w}(t) + qA_{p,q-1,\cdots w}(t) + \cdots$$
$$+ wA_{p,q,\cdots,w-1}(t)$$

In particular

$$pA_{pq}(t) = [q + 1 - p + t(2p-1)]A_{p-1,q}(t)$$
$$+ (1-t)\,DA_{p-1,q} - (p-1)(1-t)^2A_{p-2,q}$$
$$DA_{pq}(t) = pA_{p-1,q}(t) + qA_{p,q-1}(t)$$
$$pA_{pqr}(t) = [q + r + 1 - p + t(2p-1)]A_{p-1,q,r}(t) + (1-t)\,DA_{p-1,qr}(t)$$
$$- (p-1)(1-t)^2A_{p-2,qr}$$
$$DA_{pqr}(t) = pA_{p-1,qr} + qA_{p,q-1,r}(t) + rA_{pq,r-1}(t)$$

19. For two decks each of specification $ppp$, use the relations of Problem 18 to show that, writing $B_p$ for $A_{ppp}(t)$,

$$3p^2B_p - [p + 1 + (2p-1)t]\,DB_p - t(1-t)\,D^2B_p$$
$$= 2p(1-t)^2[(3p - 3 + (6p-3)t)B_{p-1} + (1-t)(2+t)\,DB_{p-1}]$$
$$2p\,DB_p = [3(3p^2-1) + 6(3p-1)(p-1)t + 3(9p^2 - 10p + 3)t^2]B_{p-1}$$
$$+ [7p - 4 + (12p-9)t - (9p-6)t^2 - (10p-7)t^3]\,DB_{p-1}$$
$$+ (2+t)(1+t)(1-t)^2\,D^2B_{p-1}$$

20. Using the match matrix (for two decks each of specification $ppp$)

$$\begin{matrix} i & j & p-i-j \\ k & l & p-k-l \\ p-i-k & p-j-l & -2p+i+j+k+l \end{matrix}$$

show that the first coefficients $B_{pk}$ in the expression

$$B_p = \Sigma B_{pk} t^k$$

are as follows

$$B_{p0} = \Sigma \binom{p}{k}^3 \qquad \text{(MacMahon)}$$

$$B_{p1} = 3p \Sigma \binom{p}{k+1}\binom{p}{k}\binom{p-1}{k}$$

$$B_{p2} = 3\binom{p}{2} \Sigma \binom{p}{k+2}\binom{p}{k}\binom{p-2}{k} + 3p^2 \Sigma \binom{p}{k+1}\binom{p-1}{k+1}\binom{p-1}{k}$$

$$B_{p3} = 3\binom{p}{3} \Sigma \binom{p}{k+3}\binom{p}{k}\binom{p-3}{k} + 6p\binom{p}{2}\Sigma\binom{p}{k+1}\binom{p-1}{k+2}\binom{p-2}{k} + p^3 \Sigma \binom{p-1}{k}^3$$

Verify the following table

| $P$ | 1 | 2 | 3 | 4 | 5 |
|---|---|---|---|---|---|
| $B_{p0}$ | 2 | 10 | 56 | 346 | 2252 |
| $B_{p1}$ | 3 | 24 | 216 | 1824 | 15150 |
| $B_{p2}$ | 0 | 27 | 378 | 4536 | 48600 |
| $B_{p3}$ | 1 | 16 | 435 | 7136 | 99350 |

TABLE 1

DISTINCT ROOK POLYNOMIALS FOR CONNECTED BOARDS WITH $n$ CELLS

| | Polynomial | Board |
|---|---|---|
| $n = 1$ | $1 + x$ | × |
| $n = 2$ | $1 + 2x$ | ×× |
| $n = 3$ | $1 + 3x$ | ××× |
| | $1 + 3x + x^2$ | ××<br>  × |
| $n = 4$ | $1 + 4x$ | ×××× |
| | $1 + 4x + 2x^2$ | ××× ∼ ××<br>    ×     ×× |
| | $1 + 4x + 3x^2$ | ××    ∼ ×××<br>  ××        × |
| $n = 5$ | $1 + 5x$ | ××××× |
| | $1 + 5x + 3x^2$ | ××××<br>      × |

TABLE 1 (*continued*)

| | Polynomial | Board |
|---|---|---|
| $n = 5$ | $1 + 5x + 4x^2$ | × × × ∼ × × × ∼ × × × ×<br>× ×  ×  ×<br>× |
| | $1 + 5x + 5x^2$ | × × ×<br>× × |
| | $1 + 5x + 5x^2 + x^3$ | × × ×<br>×<br>× |
| | $1 + 5x + 6x^2 + x^3$ | × ×<br>× ×<br>× |
| $n = 6$ | $1 + 6x$ | × × × × × × |
| | $1 + 6x + 4x^2$ | × × × × ×<br>× |
| | $1 + 6x + 6x^2$ | × × × × ∼ × × × × ∼ × × ×<br>× ×  ×  × × ×<br>× |
| | $1 + 6x + 7x^2$ | × × × × ∼ × × ×<br>× ×  × × × |
| | $1 + 6x + 8x^2$ | × × × ∼ × × × ×<br>× × ×  × × |
| | $1 + 6x + 7x^2 + x^3$ | × × ×<br>× ×<br>× |
| | $1 + 6x + 7x^2 + 2x^3$ | ×<br>× × × ×<br>× |
| | $1 + 6x + 8x^2 + 2x^3$ | × × ∼ ×<br>× × ×  ×<br>×  × × ×<br>× |
| | $1 + 6x + 9x^2 + 2x^3$ | × × ∼ × × ×<br>× ×  × ×<br>×  ×  × |
| | $1 + 6x + 9x^2 + 3x^3$ | × ×<br>× × ×<br>× |
| | $1 + 6x + 9x^2 + 4x^3$ | × × × ∼ × × ×<br>×  × ×<br>×  ×<br>× |
| | $1 + 6x + 10x^2 + 4x^3$ | × ×<br>× ×<br>× × |

CHAPTER 8

# Permutations with Restricted Position II

## 1. INTRODUCTION

In this chapter the discussion of the topics introduced in Chapter 7 is continued, moving on to the staircase chessboards, the most famous example of which is the ménage problem, to Latin rectangles which are closely associated, and, finally, to the trapezoidal and triangular boards which appear in Simon Newcomb's problem. An interesting use of the last is in the recreation known as the problem of the bishops. Each of these topics has a growing end and the treatment of the text and its continuation in the problems merely serve to define the open regions; a striking example is that of Latin rectangles where the enumeration for more than three lines has scarcely been begun.

## 2. THE MÉNAGE PROBLEM

This has been stated in Section 7.1, in the formulation given by E. Lucas, **16**, in 1891; the reduced problem is that of enumerating permutations of elements 1 to $n$ such that element $i$ is in neither of positions $i$ and $i + 1$, $i = 1$ to $n - 1$, and element $n$ in neither of positions $n$ and 1. Put in another way, it requires the enumeration of permutations discordant with the two permutations $123 \cdots n$ and $234 \cdots 1$.

Thirteen years before Lucas, Cayley, **6**, and Muir, **17**, had examined this second form of the problem, following a suggestion from P. G. Tait, **34**, who believed he needed the enumeration in his study of knots. (Later, it developed that Tait's problem was harder, namely, the enumeration of ménage permutations when those related by $M_1 = CM_2C^{-1}$, with $M_1$ and $M_2$ ménage permutations and $C$ a cyclic permutation, are regarded as the same, but no solution was given.)* Cayley and Muir, like Lucas, found

* Reading this remark in typescript has led E. N. Gilbert to produce a solution which is given in "Knots and classes of ménage permutations", *Scripta Math.*, vol. 22 (1956), pp. 228–233.

recurrence relations but missed the simple expression, equation (5) below, found later by Touchard, **35**.

It is convenient to develop together the solutions for the straight and circular table versions of the ménage problem. The board for the first is the staircase

$$\begin{array}{lllllll} \times & & & & & & \\ \times & \times & & & & & \\ & \times & \cdots & & & & \\ & & & & \times & \times & \\ & & & & & \times & \times \end{array}$$

having $2n - 1$ cells, two in each of $n$ rows except the first, and two in each of $n$ columns except the last.

The board for the circular table is the same with the addition of a cell in the first row and last column.

Let $L_k(x)$ be the rook polynomial for the first $k$ cells of the staircase shown (reading down and to the right); $L_{2n-1}(x)$ is the polynomial for the straight table. Let $M_n(x)$ be the polynomial for the circular table.

Then, by expansion with respect to the single cell in the first row of the staircase with $k$ cells

$$L_k(x) = L_{k-1}(x) + xL_{k-2}(x) \tag{1}$$

an equation which has appeared already, equation (7.26), as that of the "largest" rook polynomial with $k$ cells. Then, just as before, equation (7.27),

$$L_k(x) = \sum_0^m \binom{k-j+1}{j} x^j, \qquad m = [(k+1)/2] \tag{2}$$

and the rook polynomial for the straight board is

$$L_{2n-1}(x) = \sum_0^n \binom{2n-k}{k} x^k \tag{2a}$$

For the ménage rook polynomial $M_n(x)$, by expansion with respect to the added cell, it follows at once that

$$M_n(x) = L_{2n-1}(x) + xL_{2n-3}(x), \qquad n > 1 \tag{3}$$

hence, by equation (2),

$$M_n(x) = \sum \frac{2n}{2n-k} \binom{2n-k}{k} x^k, \qquad n > 1 \tag{4}$$

The hit polynomials may now be written down at once from equation (7.1), and are summarized in:

**Theorem 1.** *The hit polynomial* $U_n(t)$ *for the reduced ménage problem and its correspondent* $V_n(t)$ *for the straight table are given by*

$$U_n(t) = \sum_0^n \frac{2n}{2n-k}\binom{2n-k}{k}(n-k)!\,(t-1)^k \tag{5}$$

$$V_n(t) = \sum_0^n \binom{2n-k}{k}(n-k)!\,(t-1)^k \tag{6}$$

*and*

$$U_n(t) = V_n(t) + (t-1)V_{n-1}(t) \tag{7}$$

*The rook polynomials are given respectively by equations* (4) *and* (2*a*).

Equation (5) is due, essentially, to Touchard, **35**.

Note that equation (7) follows from (3). Note also that $U_n(t)$ is defined only for $n$ greater than one and $V_n(t)$ only for $n$ greater than zero. However, by equation (4), $M_1(x) = 1 + 2x$, which implies $U_2(t) = -1 + 2t$, a natural value; the value for $n = 0$ is set by convention and it is sometimes convenient, as will appear, to take $M_0 = U_0 = 2$ rather than the usual $M_0 = U_0 = 1$. For the straight table, on the other hand, the natural choice is $L_0 = V_0 = 1$.

Table 1 shows the polynomial $U_n(t)$ for $n = 2(1)10$; Table 2 shows $V_n(t)$ for the same range.

TABLE 1

COEFFICIENTS OF MÉNAGE HIT POLYNOMIAL $U_n(t)$

$U_n(t) = \Sigma U_{nk}t^k$

| $k\backslash n$ | 2 | 3 | 4 | 5 | 6 | 7 | 8 | 9 | 10 |
|---|---|---|---|---|---|---|---|---|---|
| 0 | 0 | 1 | 2 | 13 | 80 | 579 | 4738 | 43387 | 439792 |
| 1 | 0 | 0 | 8 | 30 | 192 | 1344 | 10800 | 97434 | 976000 |
| 2 | 2 | 3 | 4 | 40 | 210 | 1477 | 11672 | 104256 | 1036050 |
| 3 | | 2 | 8 | 20 | 152 | 994 | 7888 | 70152 | 695760 |
| 4 | | | 2 | 15 | 60 | 469 | 3660 | 32958 | 328920 |
| 5 | | | | 2 | 24 | 140 | 1232 | 11268 | 115056 |
| 6 | | | | | 2 | 35 | 280 | 2856 | 30300 |
| 7 | | | | | | 2 | 48 | 504 | 6000 |
| 8 | | | | | | | 2 | 63 | 840 |
| 9 | | | | | | | | 2 | 80 |
| 10 | | | | | | | | | 2 |

Though Theorem 1 contains the complete solution of the ménage problem, an extensive development of its consequences is required, both for numerical purposes and for use in related problems. To examine some of these, it is worth-while looking at another derivation, actually the one constituting the first published proof (Kaplansky, **10**).

TABLE 2

COEFFICIENTS OF MÉNAGE HIT POLYNOMIAL $V_n(t)$

$$V_n(t) = \Sigma V_{nk} t^k$$

| $k \backslash n$ | 1 | 2 | 3 | 4 | 5 | 6 | 7 | 8 | 9 | 10 |
|---|---|---|---|---|---|---|---|---|---|---|
| 0 | 0 | 0 | 1 | 3 | 16 | 96 | 675 | 5413 | 48800 | 488592 |
| 1 | 1 | 1 | 1 | 8 | 35 | 211 | 1459 | 11584 | 103605 | 1030805 |
| 2 | | 1 | 3 | 6 | 38 | 213 | 1479 | 11692 | 104364 | 1036809 |
| 3 | | | 1 | 6 | 20 | 134 | 915 | 7324 | 65784 | 657180 |
| 4 | | | | 1 | 10 | 50 | 385 | 3130 | 28764 | 291900 |
| 5 | | | | | 1 | 15 | 105 | 952 | 9090 | 95382 |
| 6 | | | | | | 1 | 21 | 196 | 2100 | 23310 |
| 7 | | | | | | | 1 | 28 | 336 | 4236 |
| 8 | | | | | | | | 1 | 36 | 540 |
| 9 | | | | | | | | | 1 | 45 |
| 10 | | | | | | | | | | 1 |

The cells of the board for the straight table may be labeled (1, 1), (2, 1), (2, 2), $\cdots$, $(i, i-1)$, $(i, i)$, $\cdots$, $(n, n)$, $(i, j)$ being the label of the cell in row $i$ and column $j$. The additional cell for the circular table is $(1, n)$. Then all cells may be numbered serially by the rules $(i, j) = C_{i+j-1}$, $(1, n) = C_{2n}$. So numbered, they form sets such that each adjacent pair is incompatible from the standpoint of the rook problem, including, for the circular table, the pair $(C_{2n}, C_1)$.

Hence the coefficients of the rook polynomials may be obtained from the following lemma (Kaplansky, l.c.):

**Lemma.** *The number of ways of selecting $k$ objects, no two consecutive, from $n$ objects arrayed in a line is $\binom{n-k+1}{k}$; if the $n$ objects are arrayed on a circle, the number is $\binom{n-k}{k} n/(n-k)$.*

To prove this, first take $f(n, k)$ for the first number. The corresponding selections either include the first object or they do not. If they do, they cannot include the second object and, hence, are enumerated by $f(n-2, k-1)$; if they do not, they are enumerated by $f(n-1, k)$. Hence,

$$f(n, k) = f(n-1, k) + f(n-2, k-1)$$

The boundary conditions are $f(n, 1) = n$, $f(1, n) = 0$, $n > 1$, which clearly establish the result in the lemma.

Next, take $g(n, k)$ for the second number, the number of choices for objects arrayed in a circle. Then, as above, the selections either contain the first object or they do not. If they do, the enumerator is now $f(n-3, k-1)$ since neither the second nor the last object may be included; if they do not, the enumerator is $f(n-1, k)$ as above. Hence,

$$g(n, k) = f(n-1, k) + f(n-3, k-1)$$
$$= \binom{n-k}{k} + \binom{n-k-1}{k-1} = \frac{n}{n-k}\binom{n-k}{k}$$
$$= \frac{n}{k}\binom{n-k-1}{k-1}$$

as stated.

It may be noted that the two generating functions

$$f_n(x) = \sum_0^m f(n, k)x^k, \qquad m = [(n+1)/2]$$

$$g_n(x) = \sum_0^q g(n, k)x^k, \qquad q = [n/2]$$

have the same recurrence, equation (1), as $L_n(x)$. Indeed it is clear that $f_n(x) \equiv L_n(x)$, while $g_{2n}(x) \equiv M_n(x)$. Since

$$g_{2n}(x) = g_{2n-1}(x) + xg_{2n-2}(x)$$
$$= (1 + x)g_{2n-2}(x) + xg_{2n-3}(x)$$
$$= (1 + 2x)g_{2n-2}(x) - x^2g_{2n-4}(x)$$

it follows that

$$M_n(x) = (1 + 2x)M_{n-1}(x) - x^2M_{n-2}(x) \tag{8}$$

which is consonant with equation (4), if $M_0(x) = 2$, $M_1(x) = 1 + 2x$. Similarly,

$$L_{2n-1}(x) = (1 + 2x)L_{2n-3}(x) - x^2L_{2n-5}(x) \tag{9}$$

which is, of course, identical in form with (8).

Noting that (8) implies

$$M_{n,k} = M_{n-1,k} + 2M_{n-1,k-1} - M_{n-2,k-2}$$

leads at once to the recurrence

$$U_n(t) = \sum_0^n (M_{n-1,k} + 2M_{n-1,k-1} - M_{n-2,k-2})(n-k)!\,(t-1)^k$$
$$= (n - 2 + 2t)U_{n-1}(t) - (t-1)U'_{n-1}(t) - (t-1)^2U_{n-2}(t) \tag{10}$$

the prime denoting a derivative. The corresponding relation for $V_n(t)$ is identical in form and will not be written out.

Somewhat simpler recurrences follow from

$$\begin{aligned} M_{nk} &= \frac{2n}{2n-k}\binom{2n-k}{k} = \frac{2n}{k}\binom{2n-1-k}{k-1} \\ &= \frac{2n}{2n-2}\,\frac{2n-1-k}{k}\,\frac{2n-2}{k-1}\binom{2n-2-k}{k-2} \\ &= \frac{n}{n-1}\,\frac{2n-1-k}{k}\,M_{n-1,k-1} \end{aligned}$$

which corresponds to the polynomial relation (primes denote derivatives)

$$(n-1)M_n'(x) = 2n(n-1)M_{n-1}(x) - nxM_{n-1}'(x)$$

and

$$L_{2n-1,k} = \frac{2n-k}{k}\,L_{2n-3,k-1}$$

which corresponds to

$$L_{2n-1}'(x) = (2n-1)L_{2n-3}(x) - xL_{2n-3}'(x)$$

These imply

$$(n-1)U_n'(t) = 2n(n-1)U_{n-1}(t) + n(1-t)U_{n-1}'(t) \qquad (11)$$

$$V_n'(t) = (2n-1)V_{n-1}(t) + (1-t)V_{n-1}'(t) \qquad (12)$$

The corresponding coefficient relations are

$$(n-1)kU_{n,k} = 2(2n-1-k)U_{n-1,k-1} + nkU_{n-1,k} \qquad (13)$$

$$kV_{n,k} = (2n-k)V_{n-1,k-1} + kV_{n-1,k} \qquad (14)$$

Also, since

$$M_{n,k} = (2n/k)L_{2n-3,k-1}$$

then

$$\begin{aligned} U_n'(t) &= 2nV_{n-1}(t) \qquad (15) \\ kU_{n,k} &= 2nV_{n-1,k-1} \end{aligned}$$

Further recurrences not involving derivatives are obtained by use of an auxiliary polynomial (which seems to have no combinatorial meaning: it is formally the hit polynomial corresponding to the rook polynomial $L_{2n}(x)$ in a square of side $n$), namely,

$$W_n(t) = \Sigma \binom{2n-k+1}{k}(n-k)!\,(t-1)^k \qquad (16)$$

Then

$$\begin{aligned} U_n(t) &= nW_{n-1}(t) + 2(t-1)^n \\ &= W_n(t) - (t-1)^2W_{n-2}(t) \qquad (17) \end{aligned}$$

the last by use of

$$M_{n,k} = \binom{2n-k+1}{k} - \binom{2n-k-1}{k-2}$$

Hence,

$$W_n(t) = nW_{n-1}(t) + (t-1)^2 W_{n-2}(t) + 2(t-1)^n$$

and, using the first of equations (17) in this,

$$(n-2)U_n(t) = n(n-2)U_{n-1}(t) + n(t-1)^2 U_{n-2}(t) - 4(t-1)^n \quad (18)$$

The instance of (18) for $t = 0$, writing $U_{n,0} = U_n$, is

$$(n-2)U_n = n(n-2)U_{n-1} + nU_{n-2} + 4(-1)^{n+1} \quad (19)$$

a result which is due (essentially) to Cayley, **6**.

Since, by the binomial coefficient recurrence

$$V_n(t) = W_n(t) - (t-1)W_{n-1}(t)$$

and, by (7) and (17),

$$V_n(t) + (t-1)V_{n-1}(t) = U_n(t) = nW_{n-1}(t) + 2(t-1)^n$$

it is found in a similar way that

$$(n-1)V_n(t) = (n^2 - n - 1 + t)V_{n-1}(t) + n(t-1)^2 V_{n-2}(t) \\ -2(t-1)^n \quad (20)$$

and, with $V_n(0) \equiv V_n$,

$$(n-1)V_n = (n^2 - n - 1)V_{n-1} + nV_{n-2} + 2(-1)^{n+1} \quad (21)$$

Further properties of the various numbers and polynomials above appear in the problems.

## 3. PERMUTATIONS DISCORDANT WITH TWO GIVEN PERMUTATIONS

As has been noticed above, the ménage numbers $U_n = U_n(0)$ enumerate permutations discordant with the two permutations $12 \cdots n$ and $23 \cdots n1$. Indeed, in Touchard's study, **35**, in which the result in equation (5) first appears, they are incidental to the enumeration of permutations discordant with any two permutations; this work is summarized in present terms below.

As before, the problem is completely determined by the character of the chessboard corresponding to the given permutations, or, since one may be given the standard order, by the (relative) cycle structure of the other. In the standard order, the first corresponds to cells on the diagonal of a square. The unit cycles of the second add no cells; the cycles of $k$ add cells which join with those on the diagonal to produce a board having rook polynomial $M_k(x)$. The boards for separate cycles are disjunct. So, for a permutation with cycle structure $1^{k_1}2^{k_2}\cdots n^{k_n}$, or for brevity of cycle class $(k)$, with, of course, $k_1 + 2k_2 + \cdots + nk_n = n$, the rook polynomial is

$$R(x, (k)) = (1 + x)^{k_1}(M_2(x))^{k_2}\cdots(M_n(x))^{k_n} \tag{22}$$

Here it is convenient to deal with the associated rook polynomial

$$\begin{aligned} r(x, (k)) &= x^n R(-x^{-1}, (k)) \\ &= (x - 1)^{k_1}(m_2(x))^{k_2}\cdots(m_n(x))^{k_n} \end{aligned} \tag{22a}$$

with

$$m_k(x) = x^k M_k(-x^{-1})$$

The corresponding hit polynomial, by equation (7.3$a$), may be written as

$$U(t, (k)) = (1 - t)^n r[E(1 - t)^{-1}, (k)]0! \tag{23}$$

and if $r(x, (k))$ may be expressed as a linear sum of the ménage associated rook polynomials, $m_k(x)$, the hit polynomial, by (23), will be a sum of the ménage hit polynomials, $U_k(t)$.

To show this, note first that by equation (8)

$$m_n(x) = (x - 2)m_{n-1}(x) - m_{n-2}(x), \qquad n > 1$$

and, as will appear, it is convenient to take $m_0(x) = 2$, $m_1(x) = x - 2$. This recurrence may be compared with that for Chebyshev polynomials

$$T_n(x) = \cos n \arccos x$$

(so that $T_0 = 1$, $T_1 = x$, $T_2 = 2x^2 - 1$, and so on), which is

$$T_n(x) = 2xT_{n-1}(x) - T_{n-2}(x)$$

Then

$$\begin{aligned} T_{2n}(x) &= 2xT_{2n-1}(x) - T_{2n-2}(x) \\ &= (4x^2 - 1)T_{2n-2}(x) - 2xT_{2n-3}(x) \\ &= (4x^2 - 2)T_{2n-2}(x) - T_{2n-4}(x) \end{aligned}$$

Hence

$$\begin{aligned} m_n(x) &= 2T_{2n}(\sqrt{x}/2) \\ &= 2\cos 2n\theta, \qquad \cos\theta = \sqrt{x}/2 \end{aligned} \tag{24}$$

Note that this gives $m_0 = 2$, $m_1 = x - 2$ as above.

This has the immediate consequence, with $\theta$ as above:

$$\begin{aligned} m_j(x)m_k(x) &= 4\cos 2j\theta \cos 2k\theta \\ &= 2[\cos 2(j + k)\theta + \cos 2(j - k)\theta] \\ &= m_{j+k}(x) + m_{j-k}(x) \end{aligned} \tag{25}$$

if, as is natural with the even functions involved, the convention $m_{-n}(x) \equiv m_n(x)$ is followed. In particular

$$(m_j(x))^2 = m_{2j}(x) + m_0(x)$$

Next, iteration of (25) leads to, omitting the variable $x$ for brevity,

$$m_i m_j m_k = m_{i+j+k} + m_{i+j-k} + m_{i-j+k} + m_{i-j-k}$$
$$m_i^3 = m_{3i} + 3m_i$$

and it is clear that the product of $k$ polynomials may be written as a sum of $2^{k-1}$ terms as in

$$m_{i_1} m_{i_2} \cdots m_{i_k} = \Sigma m_{i_1 \pm i_2 \pm \cdots \pm i_k} \tag{26}$$

Since this holds also when some or all of the $i_j$ are alike, it follows that any expression $m_2^{k_2} m_3^{k_3} \cdots$ may be reduced to a sum of associated ménage polynomials. On the other hand, for a product like $(x - 1)^{k_1} m_n$, notice first that by equation (8)

$$(x - 1)m_n = m_{n+1} + m_n + m_{n-1}$$

and

$$(x - 1)^2 m_n = (x - 1)m_{n+1} + (x - 1)m_n + (x - 1)m_{n-1}$$
$$= m_{n+2} + 2m_{n+1} + 3m_n + 2m_{n-1} + m_{n-2}$$

The general formula is simply expressed in the symbolic form

$$(x - 1)^k m_n(x) = m^{n-k}(1 + m + m^2)^k, \qquad m^j \equiv m_j(x) \tag{27}$$

Thus, finally, any polynomial of the form (22*a*) may be expressed as a linear sum of associated ménage polynomials, and the results of this section may be summarized in

**Theorem 2.** *The permutations discordant with two permutations having a relative cycle structure of class* $(k) = (k_1, k_2, \cdots)$ *correspond to associated rook polynomial*

$$r(x, (k)) = (x - 1)^{k_1}(m_2(x))^{k_2} \cdots$$

*which, by equations* (26) *and* (27), *may be reduced to*

$$r(x, k)) = \Sigma A_j m_{n-j}(x) \tag{22b}$$

*with* $A_j$ *a function of the cycle class* $(k)$ *and* $n = k_1 + 2k_2 + \cdots$. *The corresponding hit polynomial is then given by*

$$U(t, (k)) = \Sigma A_j(1 - t)^j U_{n-j}(t) \tag{23a}$$

*with* $U_n(t)$ *the ménage hit polynomial* (*equation* (5)).

Formal expressions for specific classes are too involved to be written out in any but the simplest cases illustrated by

*Example 1.* For permutations discordant with 123 and 132, the associated rook polynomial is $(x - 1)m_2 = x^3 - 5x^2 + 6x - 2$, and the hit polynomial is $4t + 2t^3$. By equation (27)

$$(x - 1)m_2 = m_3 + m_2 + m_1$$

and by equation (23) the hit polynomial is $U_3(t) + (1 - t)U_2(t) + (1 - t)^2U_1(t)$, which is equal to $4t + 2t^3$, as above. More generally, for permutations discordant with $12 \cdots n$ and $134 \cdots n2$, the rook polynomial is $(x - 1)m_{n-1} = m^{n-2}(1 + m + m^2)$, $m^k \equiv m_k$, and the hit polynomial is

$$U_n(t) + (1 - t)U_{n-1}(t) + (1 - t)^2U_{n-2}(t) = U^{n-2}[U^2 + (1 - t)U + (1 - t)^2], \qquad U^n \equiv U_n(t)$$

Finally, the hit polynomial corresponding to rook polynomial $(x - 1)^k m_n(x)$ by equations (27) and (23) is

$$U^{n-k}[U^2 + (1 - t)U + (1 - t)^2]^k, \qquad U^n \equiv U_n(t)$$

## 4. LATIN RECTANGLES

As already mentioned in Section 7.1, a Latin rectangle of $k$ rows and $n$ columns contains a permutation of elements 1 to $n$ in each row, these permutations being so chosen that no column contains repeated elements. The simplest 3 by $n$ Latin rectangle is

$$\begin{matrix} 1 & 2 & \cdots & n \\ 2 & 3 & \cdots & 1 \\ 3 & 4 & \cdots & 2 \end{matrix}$$

When the first row is written in the natural order, the rectangle is said to be reduced. If $L(k, n)$ is the number of $k$ by $n$ Latin rectangles and $K(k, n)$ the number of reduced rectangles, $L(k, n) = n!\,K(k, n)$. As already noted, $K(2, n) = D_n$, the displacement number of Chapter 3, and $K(3, n)$ may be expressed by the ménage number $U_n$, as will now be shown.

In the language of the preceding section, $K(3, n)$ enumerates permutations discordant with every choice of two permutations, one of which is $12 \cdots n$, and the other a permutation discordant with it. Hence, the second permutation has a cycle structure $(2^{k_2}3^{k_3} \cdots n^{k_n})$ with $2k_2 + 3k_3 + \cdots + nk_n = n$, that is, a structure without unit parts (which would permit like numbers in the same column). By Theorem 2, the permutations discordant with two such permutations are enumerated by

$$U(0, (k)) = \Sigma A_j U_{n-j}$$

with $U_k \equiv U_k(0)$ a ménage number and

$$m_2^{k_2} m_3^{k_3} \cdots m_n^{k_n} = \Sigma A_j m_{n-j}, \qquad 2k_2 + 3k_3 + \cdots + nk_n = n$$

Hence,

$$K(3, n) = \sum_{j} U_{n-j} \sum_{(k)} A_j[(k)] \tag{28}$$

with $(k)$ any non-unitary cycle class, and $A_j[(k)]$ the number associated with the reduction of a particular $(k)$.

The cycle indicator for non-unitary permutations is $C_n(0, t_2, \cdots, t_n)$, which then gives all terms making up the inner sum in (28). Thus, since

$$C_4(0, t_2, t_3, t_4) = 3t_2^2 + 6t_4$$

and

$$m_2^2 = m_4 + m_0$$

it follows that

$$K(3, 4) = 9U_4 + 3U_0 = 24$$

The enumeration of cycle classes and their reduction can be combined into a single operation once it is noticed that $U_j$ can be used in place of $m_j$ in the reduction equation preceding (28) and that the reductions are the consequence of the rule: $U_i U_j = U_{i+j} + U_{i-j}$, with $U_{-n} = U_n$, $U_0 = 2$. Since

$$\tfrac{1}{2}(U^i + U^{-i})(U^j + U^{-j}) = U^{i+j} + U^{i-j}$$

it follows that ($K_n = K(3, n)$)

$$\begin{aligned} \exp tK &= \exp tC(0, U_2, U_3, \cdots), \qquad K^n \equiv K_n \\ &= \tfrac{1}{2} \exp tC(0, U^2 + U^{-2}, U^3 + U^{-3}, \cdots), \qquad U^n = U^{-n} \equiv U_n \\ &= \frac{1}{2} \frac{e^{-tU}}{1 - tU} \frac{e^{-t/U}}{1 - t/U} = \frac{1}{2} \frac{e^{-t\mu}}{1 - t\mu + t^2} \end{aligned} \tag{29}$$

with $\mu = U + U^{-1}$. The last line uses Problem 4.1 and equation (4.3$a$).

Since by equation (3.23)

$$\exp tD = e^{-t}(1 - t)^{-1}, \qquad D^n \equiv D_n = \Delta^n 0!$$

another form of (29) is

$$\exp tK = \tfrac{1}{2} \exp (tDU + tDU^{-1}) \tag{29a}$$

Note that in the interpretation of (29$a$), a term $D^i D^j U^i U^{-j}$ becomes $D_i D_j U_{i-j}$, and, hence,

$$K_n = \frac{1}{2} \sum_{0}^{n} \binom{n}{k} D_{n-k} D_k U_{n-2k}, \qquad U_0 = 2, \qquad U_{-n} = U_n \tag{30}$$

In particular

$$\begin{aligned} K_4 &= \tfrac{1}{2}(D_4 D_0 U_4 + 4D_3 D_1 U_2 + 6D_2^2 U_0 + 4D_1 D_3 U_{-2} + D_0 D_4 U_{-4}) \\ &= D_4 D_0 U_4 + 3D_2^2 U_0 = 24 \end{aligned}$$

as above.

As this example shows the sum is symmetric and, hence, equivalent to

$$K_n = \sum_0^m \binom{n}{k} D_{n-k} D_k U_{n-2k}, \qquad m = [n/2], \qquad U_0 = 1 \qquad (30a)$$

a result first derived in Riordan, **21**. It may be noticed that the dominant term in (30*a*) is $D_n U_n$ which is of the order of $n!^2 e^{-3}$ (the asymptotic expression for $U_n$ appears in Problem 7*b*).

Turning now to the second form of equation (29) note first that by Problem 7*a*

$$\mu^n = (U + U^{-1})^n = 2e_n$$

with $e_n = (E - 2)^n 0!$, as in Problem 7.8. Hence,

$$\begin{aligned} \exp tK &= \frac{e^{-t\mu}}{2(1 - t\mu + t^2)} \\ &= \tfrac{1}{2}\Sigma(-1)^j e^{-t\mu} t^{2j}/(1 - t\mu)^{j+1} \\ &= \tfrac{1}{2}\Sigma(-1)^j e^{-t\mu} t^{2j} \frac{(D+1)^j}{j!} \exp t\mu(D+1), \qquad D^n \equiv D_n = D^n 0! \end{aligned}$$

the relation

$$(D + 1)^j \exp t(D + 1) = j!(1 - t)^{-j-1}$$

being obtained by successive differentiation of equation (3.23). Thus, finally

$$K_n = \sum_{j=0}^m (-1)^j \frac{n!}{j!\,(n - 2j)!} (D + 1)^j D^{n-2j} e_{n-2j}, \qquad m = [n/2] \quad (31)$$

This alternative to equation (30*a*) is inferior for direct calculation because of the alternating signs of its coefficients. But it is useful in finding a recurrence for the numbers $K_n$, as will now be shown.

First notice that it is implied equally by

$$\exp tK = (1 - te + t^2)^{-1} \exp(-te), \qquad e^n \equiv e_n, \qquad K^n \equiv K_n$$

and from this it follows at once that

$$K_n = neK_{n-1} - (n)_2 K_{n-2} + (-1)^n e_n \qquad (32)$$

with $K_n$ now a function of a Blissard variable $e$. The first few instances of (32), which illustrate this functional dependence, are

$$\begin{aligned} K_0 &= e_0 = 1 \\ K_1 &= eK_0 - e_1 = e_1 - e_1 = 0 \\ K_2 &= 2eK_1 - 2K_0 + e_2 = e_2 - 2 = 0 \\ K_3 &= 3eK_2 - 6K_1 - e_3 = 3(e_3 - 2e_1) - e_3 = 2e_3 - 6e_1 = 2 \end{aligned}$$

Next, writing (31) as

$$K_n = \Sigma(-1)^j K_{n,j} e_{n-2j} \tag{31a}$$

it follows at once from (32) that

$$K_{n,0} = nK_{n-1,0} + (-1)^n = D_n$$

$$K_{n,j} = nK_{n-1,j} + (n)_2 K_{n-2,j-1}, \qquad j > 0$$

These imply

$$\begin{aligned} k_j(x) &= \Sigma K_{n,j} x^n/n! \\ &= xk_j(x) + x^2 k_{j-1}(x) \\ &= x^2(1-x)^{-1} k_{j-1}(x) = x^{2j}(1-x)^{-j-1} e^{-x} \end{aligned}$$

since $k_0(x) = \exp xD = e^{-x}(1-x)^{-1}$, by equation (3.23). Differentiation of the last equation shows, with a prime denoting a derivative, that

$$x^2 k_j'(x) = x(2j - x)k_j(x) + (j+1)k_{j+1}(x)$$

and, hence, that

$$jK_{n,j} = n(n+1-2j)K_{n-1,j-1} + (n)_2 K_{n-2,j-1}, \qquad j > 0$$

These recurrences for the coefficients $K_{n,j}$ along with

$$e_n = (n-2)e_{n-1} + 2(n-1)e_{n-2}$$

[which follows from $e_n - ne_{n-1} = (-2)^n = (-2)(e_{n-1} - (n-1)e_{n-2})$] provide a companion to equation (32) which may be used to eliminate the functional dependence. Thus, first,

$$\begin{aligned} eK_n &= \Sigma(-1)^j K_{nj} e_{n+1-2j} \\ &= \Sigma(-1)^j K_{nj}[(n-1-2j)e_{n-2j} + 2(n-2j)e_{n-1-2j}] \\ &= (n-1)K_n + I_n + J_n \end{aligned}$$

with

$$\begin{aligned} I_n &= \Sigma(-1)^{j+1} 2jK_{nj} e_{n-2j} \\ &= \Sigma(-1)^{j+1} 2e_{n-2j}[n(n+1-2j)K_{n-1,j-1} + (n)_2 K_{n-2,j-1}] \\ &= \Sigma(-1)^j 2e_{n-2-2j}[n(n-1-2j)K_{n-1,j} + (n)_2 K_{n-2,j}] \\ &= 2(n)_2 K_{n-2} + nJ_{n-1} \end{aligned}$$

and

$$\begin{aligned} J_n &= \Sigma(-1)^j 2(n-2j)e_{n-1-2j} K_{nj} \\ &= 2ne_{n-1}(nK_{n-1,0} + (-1)^n) \\ &\quad + \sum_{j=1} (-1)^j 2(n-2j)e_{n-1-2j}(nK_{n-1,j} + (n)_2 K_{n-2,j-1}) \\ &= 2n^2 K_{n-1} + (-1)^n 2ne_{n-1} + 2nI_{n-1} - (n)_2 J_{n-2} \end{aligned}$$

Next, the last equation may be reduced by its predecessor to

$$J_n = 2n^2K_{n-1} + 2(n)_3K_{n-3} + (-1)^n 2ne_{n-1} + nI_{n-1}$$

and this, in turn, by use of (32) to

$$J_n = 2(n)_2(K_{n-1} + eK_{n-2}) + nI_{n-1}$$

Hence,

$$I_n + J_n = 2(n)_2(K_{n-1} + (1+e)K_{n-2}) + n(I_{n-1} + J_{n-1})$$

or

$$(e - n + 1)K_n = 2(n)_2(K_{n-1} + (1+e)K_{n-2}) + n(e - n + 2)K_{n-1}$$

But this is the same as

$$\begin{aligned}(e - n + 1)K_n - 2(1+e)nK_{n-1} &= -n[(e - n + 2)K_{n-1} \\ &\qquad - 2(1+e)(n-1)K_{n-2}] \\ &= n(n-1)[(e - n + 3)K_{n-2} \\ &\qquad - 2(1+e)(n-2)K_{n-3}] \\ &= \cdots = (-1)^n n!(1+e)K_0 = 0 \qquad (33)\end{aligned}$$

If equation (33) is written in the form

$$neK_{n-1} - 2en(n-1)K_{n-2} = n(n-2)K_{n-1} + 2(n)_2K_{n-2}$$

and the left-hand side is reduced by (32), it is found that

$$K_n = n^2K_{n-1} + (n)_2K_{n-2} + 2(n)_3K_{n-3} + (-1)^n(e_n + 2ne_{n-1}) \qquad (34)$$

which seems to be the simplest recurrence for the numbers $K_n \equiv K(3, n)$. Its original derivation (Riordan, **25**) is quite different. A "pure" recurrence, which is much more elaborate though, as it happens, it was the first one found (Kerawala, **13**), may be obtained by eliminating $e_n$ through the recurrence $e_n = ne_{n-1} + (-2)^n$.

Equation (34) is particularly apt in finding the asymptotic series of Yamamoto, **38**, since for this purpose its "impure" term, the last term of (34), may be ignored. If the series is taken as

$$n!^{-2}\, e^3K_n \sim 1 + \frac{a_1}{n} + \frac{a_2}{(n)_2} + \cdots + \frac{a_s}{(n)_s} + \cdots$$

then from (34) and the identity

$$\frac{1}{(n-1)_s} = \frac{1}{(n)_s} + \frac{s}{(n)_{s+1}}$$

it is found that

$$a_s = a_s + (s-1)a_{s-1} + a_{s-2} + 2a_{s-3}$$

or

$$sa_s + a_{s-1} + 2a_{s-2} = 0 \qquad (35)$$

To eliminate fractions write $b_s = s!\, a_s$; then by (35),

$$b_{s+1} + b_s + 2sb_{s-1} = 0 \tag{36}$$

which implies

$$b \exp tb = -(1 + 2t) \exp tb, \qquad b^n \equiv b_n$$

The solution of this differential equation, with $b_0 = 1$, is

$$\exp tb = \exp(-t - t^2)$$

Hence, by Problem 2.25 (in the notation explained there),

$$b_n = P_n(-1, 1/2) = H_n(-1/2)$$

The first few values are

| $n$ | 0 | 1 | 2 | 3 | 4 | 5 | 6 | 7 | 8 | 9 | 10 |
|---|---|---|---|---|---|---|---|---|---|---|---|
| $b_n$ | 1 | $-1$ | $-1$ | 5 | 1 | $-41$ | 31 | 461 | $-895$ | $-6481$ | 22591 |

The results of this section may be summarized in:

**Theorem 3.** *The number of reduced 3 by n Latin rectangles, $K(3, n) \equiv K_n$, is expressed in terms of the ménage numbers by equation* (30*a*), *namely,*

$$K_n = \sum_0^m \binom{n}{k} D_{n-k} D_k U_{n-2k}, \qquad m = [n/2], \qquad U_0 = 1$$

*It has the recurrence*

$$K_n = n^2 K_{n-1} + (n)_2 K_{n-2} + 2(n)_3 K_{n-3} + (-1)^n (e_n + 2ne_{n-1})$$

*and the asymptotic series*

$$K_n \sim n!^2 e^{-3}\left(1 - \frac{1}{n} - \frac{1}{2(n)_2} + \frac{5}{6(n)_3} + \cdots + \frac{b_s}{s!(n)_s} + \cdots\right) \tag{37}$$

*with $b_s = H_s(-1/2)$.*

Table 3 shows values of $K_n$ and $e_n$ for $n = 0(1)10$, as well as values of the coefficients $K_{n,j}$ appearing in (31*a*) for the same range.

Very little is known about Latin rectangles with more than three lines. From the results

$$L(2, n) = n!\, D_n \sim n!^2\, e^{-1}$$
$$L(3, n) = n!\, K_n \sim n!^3\, e^{-3}$$

it is a natural surmise that

$$L(k, n) \sim n!^k\, e^{-\binom{k}{2}}$$

and Erdös and Kaplansky, **8**, have shown that this is true for $k < (\log n)^{3/2}$ and guessed that $k$ had the larger range $k < n^{1/3}$, which was proved later by Yamamoto, **41**.

TABLE 3

THREE-LINE LATIN RECTANGLES

NUMBERS $K_{nj}$, $e_n$, AND $K_n$; $K_n = \Sigma(-1)^j K_{n,j} e_{n-2j}$

| $j \backslash n$ | 0 | 1 | 2 | 3 | 4 | 5 | 6 | 7 | 8 | 9 |
|---|---|---|---|---|---|---|---|---|---|---|
| 0 | 1 | 0 | 1 | 2 | 9 | 44 | 265 | 1854 | 14833 | 133496 |
| 1 | | | 2 | 6 | 36 | 220 | 1590 | 12978 | 118664 | 1201464 |
| 2 | | | | | 24 | 240 | 2520 | 26880 | 304080 | 3671136 |
| 3 | | | | | | | 720 | 15120 | 262080 | 4294080 |
| 4 | | | | | | | | | 40320 | 1451520 |

| $n$ | 0 | 1 | 2 | 3 | 4 | 5 | 6 | 7 | 8 | 9 | 10 |
|---|---|---|---|---|---|---|---|---|---|---|---|
| $e_n$ | 1 | −1 | 2 | −2 | 8 | 8 | 112 | 656 | 5504 | 49024 | 4 91264 |
| $K_n$ | 1 | 0 | 0 | 2 | 24 | 552 | 21280 | 1073760 | 70299264 | 57928 53248 | 58 71599 44704 |

The enumeration of Latin squares has been carried up to 7 by 7 with the following results. Write

$$L(n, n) = n!(n-1)!\, l_n$$

Then $l_n$ is the number of squares with the first row and first column in standard order; its values are

| $n$ | 2 | 3 | 4 | 5 | 6 | 7 |
|---|---|---|---|---|---|---|
| $l_n$ | 1 | 1 | 4 | 56 | 9408 | 16942080 |

The number $l_7$ is taken from Sade, **27**.

## 5. TRAPEZOIDS AND TRIANGLES

Consider first the trapezoid board with $q$ rows, namely,

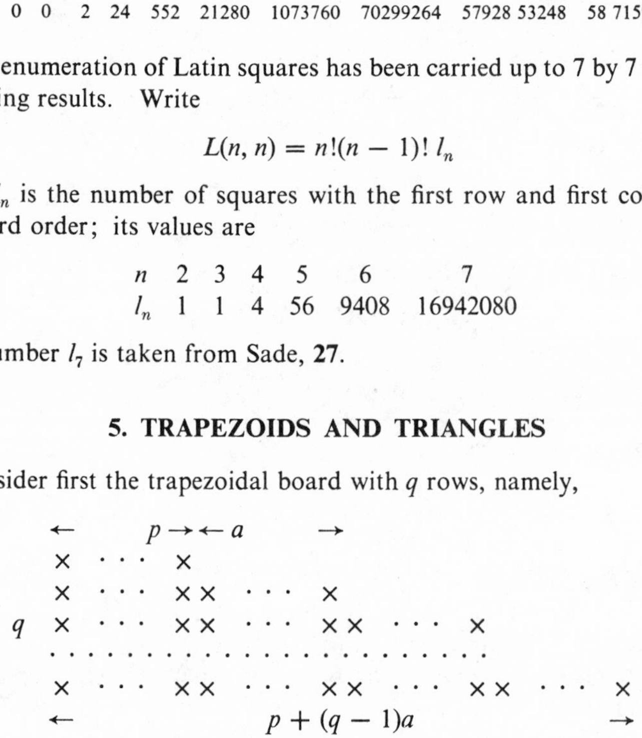

```
      ←     p →← a     →
      ×  · · ·  ×
      ×  · · ·  ××  · · ·  ×
  q   ×  · · ·  ××  · · ·  ××  · · ·  ×
      · · · · · · · · · · · · · · · · · · · ·
      ×  · · ·  ××  · · ·  ××  · · ·  ××  · · ·  ×
      ←              p + (q − 1)a              →
```

There are $p$ cells in the first row, $p + a$ in the second, and each succeeding row has $a$ more, so that finally the $q$th has $p + (q - 1)a$. The triangle is the special case $p = a = 1$.

Expanding with respect to the cells of the first row gives a basic recurrence. If $T(p, q, a; x)$ is the rook polynomial for a trapezoid as shown, this is

$$T(p, q, a; x) = T(p + a, q - 1, a; x) + pxT(p + a - 1, q - 1, a; x) \quad (38)$$

which with $T(p, 1, a; x) = 1 + px$ completely determines all trapezoid polynomials. Thus

$$T(p, 2, a; x) = 1 + x(2p + a) + x^2p(p + a - 1)$$
$$T(p, 3, a; x) = 1 + x(3p + 3a) + x^2[3p(p - 1) + 6ap + 2a^2 - a] + x^3p(p + a - 1)(p + 2a - 2)$$

Writing

$$T(p, q, a; x) = \sum_{k=0} T_k(p, q, a)x^k$$

it may be noticed that

$$T_1(p, q, a) = pq + a\binom{q}{2}$$

$$T_2(p, q, a) = [(p)_2 + ap(q - 1)]\binom{q}{2} + [(3q - 1)a^2 - 4a]\binom{q}{3}/4$$

$$T_q(p, q, a) = p(p + a - 1)(p + 2a - 2)\cdots(p + (q - 1)(a - 1))$$

and

$$T_k(p, q, a) = T_k(p + a, q - 1, a) + pT_{k-1}(p + a - 1, q - 1, a)$$

There seems to be no simple general formula.

However, for $a = 2$, the formulas above reduce to

$$T_1(p, q, 2) = q(p + q - 1)$$

$$T_2(p, q, 2) = \binom{q}{2}(p + q - 1)_2$$

$$T_q(p, q, 2) = (p + q - 1)_q$$

which suggest that

$$T_k(p, q, 2) = \binom{q}{k}(p + q - 1)_k$$

which may be proved by the recurrence relation and mathematical induction. Then

$$T(p, q, 2; x) = \Sigma\binom{q}{k}(p + q - 1)_k x^k = R_{p+q-1,q}(x) \tag{39}$$

with $R_{n,m}(x)$ the rectangular rook polynomial. Hence, for the problem of the rooks, *the trapezoid* $(p, q, 2)$ *is equivalent to a* $q$ *by* $p + q - 1$ *rectangle*. This result may also be obtained by repeated application of Theorem 7.3.

For $a = 1$, the case next in difficulty and interest, write $T_{p,q}(x) = T(p, q, 1; x)$ and $T_{1,q}(x) = T_q(x)$, the rook polynomial for the triangle. From the general formulas above, or by direct calculation, it appears that

$$T_1(x) = 1 + x$$
$$T_2(x) = 1 + 3x + x^2$$
$$T_3(x) = 1 + 6x + 7x^2 + x^3$$

and it will be noticed that the coefficients are Stirling numbers of the second kind, a result which will be proved now.

First by Problem 7.4, all cells on the main diagonal of the triangular board are equivalent for expansion; however, instead of the complete triangle it is more convenient to expand with respect to the $q$ diagonal cells of the board with polynomial $T_{2,q}(x)$ which, by (36), is equal to $T_{q+1}(x) - xT_q(x)$. These cells are disjunct and have rook polynomial $(1 + x)^q$; moreover, the board corresponding to inclusion of $k$ of these cells is a triangle with polynomial $T_{q-k}(x)$. Hence,

$$T_{q+1}(x) - xT_q(x) = (T + x)^q, \qquad T^k \equiv T_k(x) \tag{40}$$

which is in agreement with the results given above.

Writing

$$T(x, y) = \sum_{q=0}^{\infty} T_q(x)y^q/q!$$

and $T_y(x, y)$ as the partial derivative of $T(x, y)$ with respect to $y$, it follows at once from (40) that

$$T_y(x, y) = (x + e^{xy})T(x, y) \tag{41}$$

The solution of this equation which satisfies the boundary conditions $T(x, 0) = T_0(x) = 1$, $T(0, y) = e^y$, is

$$T(x, y) = \exp(xy + x^{-1}(e^{xy} - 1)) \tag{42}$$

and using the result of Problem 2.14($a$),

$$e^{-xy}T(x, y) = \sum_{k=0}^{\infty} \frac{(e^{xy} - 1)^k}{k!\, x^k} = \sum_{q=0}^{\infty} \frac{y^q}{q!} \sum_{k=0}^{q} S(q, k)x^{q-k}$$

with $S(q, k)$ a Stirling number (second kind). Hence,

$$(T - x)^q = \sum_{k=0}^{q} S(q, k)x^{q-k}, \qquad T^k \equiv T_k(x) \tag{43}$$

But equation (41) is equivalent to

$$(T - x) \exp y(T - x) = \exp yT, \qquad T^k \equiv T_k(x)$$

Hence, finally,

$$\begin{aligned} T_q(x) &= (T - x)^{q+1}, \qquad T^k \equiv T_k(x) \\ &= \Sigma S(q + 1, q + 1 - k)x^k \end{aligned} \tag{44}$$

as was to be proved.

Equation (44) supplies a new combinatorial meaning for the Stirling numbers of the second kind, namely:

*The number of ways of putting $k$ non-attacking rooks on a (right-angled isosceles) triangle of side $q - 1$ is the Stirling number $S(q, q - k)$.*

For the polynomials $T_{p,q}(x)$, note first that by the development of equation (40)

$$T_{2,q}(x) = T^q(T - x) = (T + x)^q \qquad T^k \equiv T_k(x)$$

The first follows from the instance $p = a = 1$ of (38); the instance $a = 1$ of (38), namely,

$$T_{p+1,q}(x) = T_{p,q+1}(x) - pxT_{p,q}(x) \tag{45}$$

leads by iteration to

$$\begin{aligned} T_{p,q}(x) &= T^q(T - x)(T - 2x) \cdots (T - (p - 1)x), \qquad T^k \equiv T_k(x) \\ &= T^{q-1}(T)_p, \qquad (T)_p = T(T - x) \cdots (T - (p - 1)x) \end{aligned} \tag{46}$$

The generalization of the second form is

$$T_{p,q}(x) = (T - x + px)^q, \qquad T^k \equiv T_k(x) \tag{47}$$

which may be proved by mathematical induction after noting that (45) implies

$$\begin{aligned} T_{p+1}(x, y) &= \Sigma T_{p+1,q}(x)y^q/q! \\ &= \left(\frac{d}{dy} - px\right) T_p(x, y) \end{aligned} \tag{48}$$

and that [see the equation preceding (44)]

$$(T - x) \exp y(T - x + px) = \exp y(T + px), \qquad T^k \equiv T_k(x)$$

The results of this section may be summarized in:

**Theorem 4.** *The trapezoidal chessboard of $q$ rows having successively $p, p + a, \cdots, p + (q - 1)a$ cells, in the case $a = 2$ has a rook polynomial equal to the $p + q - 1$ by $q$ rectangular board. The polynomial of the triangle ($p = a = 1$) is*

$$T_q(x) = \Sigma S(q + 1, q + 1 - k)x^k$$

*and the polynomial of the trapezoid with $a = 1$ is*

$$T_{p,q}(x) = (T - x + px)^q = T^{q-1}(T)_{p,x}, \qquad T^k \equiv T_k(x)$$

*with $(T)_{p,x} = T(T - x) \cdots (T - (p - 1)x)$.*

## 6. TRIANGULAR PERMUTATIONS

The hit polynomial for permutations subject to restrictions corresponding to a triangular board, for brevity, will be said to enumerate triangular permutations. If the board is of side less than $n$, the permutations are further qualified as incomplete triangular permutations, as with rencontres.

These hit polynomials have a double interpretation. First, they enumerate permutations according to number of elements in the positions specified by the triangular board. But, more important for statistical applications, the restriction implied by the cell $(i, j)$ may be read as $i$ is the immediate successor of $j$; hence, if $i < j$, the enumeration is by number of "descents", that is, the number of cases where $i$ is the immediate successor of $j$ and $i$ is less than $j$. When all descents are in question, the triangle is of side $n - 1$, and the number of descents is always one less than the number of ascending runs (successions of elements each of which is greater than its predecessor). Thus, in 256413, there are two descents, 64, 41, and three ascending runs, 256, 4, and 13. The enumerator by number of ascending runs is also the enumerator by number of "readings", a reading being a scan left to right, picking up elements in natural order. This is because a permutation requiring $r$ readings has a conjugate, which has elements and position interchanged, which has $r$ ascending runs, and vice versa. Thus the conjugate of 256413 is 516423, the three readings of which are 123, 4, 56. Readings have also been called locomotions (Sade, **29**) because of the operations used in sorting cars in freight classification yards. Note that the interpretation by ascending runs is that required by Simon Newcomb's problem.

Write $A_{n,m}(t)$ for the hit polynomial corresponding to a triangle of side $n - m$, having rook polynomial

$$T_{n-m}(x) = \Sigma S(n - m + 1, k)x^{n-m+1-k}$$

Then, by equation (7.1)

$$A_{n,m}(t) = \Sigma S(n - m + 1, n - m + 1 - k)(n - k)!\,(t - 1)^k \qquad (49)$$

By the recurrence relation for Stirling numbers, equation (2.37), namely,

$$S(n + 1, k) = S(n, k - 1) + kS(n, k)$$

and the derivative relation

$$(t - 1)A'_{n,m}(t) = (n + 1)A_{n,m}(t) - \Sigma S(n - m + 1, k - m + 1)(k + 1)!\,(t - 1)^{n-k}$$

it follows that

$$\begin{aligned} A_{n+1,m}(t) &= (n + 1)tA_{n,m}(t) + m(1 - t)A_{n,m}(t) + t(1 - t)A'_{n,m}(t) \\ &= (m + (n - m + 1)t)A_{n,m}(t) + t(1 - t)A'_{n,m}(t) \qquad (50) \end{aligned}$$

Hence, if

$$A_{n,m}(t) = \Sigma A_r(n, m)t^r$$

then

$$A_r(n+1, m) = (m+r)A_r(n, m) + (n-m-r+2)A_{r-1}(n, m)$$

Notice that $A_{r-m}(n, m)$ and $A_r(n, 0)$ have the same recurrence relation, namely,

$$A_r(n+1, 0) = rA_r(n, 0) + (n-r+2)A_{r-1}(n, 0)$$

which corresponds to the instance $m = 0$ of equation (50), namely,

$$\begin{aligned} A_{n+1,0}(t) &\equiv A_{n+1}(t) \\ &= (n+1)tA_n(t) + t(1-t)A_n'(t) \end{aligned}$$

which has already appeared in Problem 2.2 for polynomials $a_n(x)$ in the Eulerian numbers. Note that boundary conditions are alike: $A_0(t) = 1$, $A_1(t) = 1 + (t-1) = t$; hence, the numbers $A_r(n, 0)$ are Eulerian numbers. From Problem 2.2, it may be noted that

$$\begin{aligned} A(t, u) &= A_0(t) + A_1(t)u + A_2(t)u^2/2! \cdots \\ &= \frac{1-t}{1-t\exp u(1-t)} \end{aligned}$$

The Eulerian numbers also appear in the polynomials $A_{n,1}(t)$; indeed,

$$tA_{n,1}(t) = A_n(t), \qquad n > 0$$

This may be shown as follows. First, $A_{1,1}(t) = 1$ follows from the general relation $A_{nn}(t) = n!$, so $tA_{1,1}(t) = t = A_1(t)$. Then, with $m = 1$, equation (50) becomes

$$A_{n+1,1}(t) = (1+nt)A_{n,1}(t) + t(1-t)A_{n,1}'(t)$$

and substitution of $tA_{n,1}(t) = A_n(t)$ gives the instance $m = 0$ of (50).

Table 4 shows the Eulerian numbers for $n = 1(1)10$.

TABLE 4

EULERIAN NUMBERS $A_{nk}$

$$A_{nk} = kA_{n-1,k} + (n-k+1)A_{n-1,k-1}$$

| $k \backslash n$ | 1 | 2 | 3 | 4 | 5 | 6 | 7 | 8 | 9 | 10 |
|---|---|---|---|---|---|---|---|---|---|---|
| 1 | 1 | 1 | 1 | 1 | 1 | 1 | 1 | 1 | 1 | 1 |
| 2 | | 1 | 4 | 11 | 26 | 57 | 120 | 247 | 502 | 1013 |
| 3 | | | 1 | 11 | 66 | 302 | 1191 | 4293 | 14608 | 47840 |
| 4 | | | | 1 | 26 | 302 | 2416 | 15619 | 88234 | 455192 |
| 5 | | | | | 1 | 57 | 1191 | 15619 | 156190 | 1310354 |
| 6 | | | | | | 1 | 120 | 4293 | 88234 | 1310354 |
| 7 | | | | | | | 1 | 247 | 14608 | 455192 |
| 8 | | | | | | | | 1 | 502 | 47840 |
| 9 | | | | | | | | | 1 | 1013 |
| 10 | | | | | | | | | | 1 |

As illustrated by the table, the asymptotic representation of the distribution $A_{n,1}(t)/n!$ is not of the modified Poisson form used above; instead, it is dominantly a normal (Laplace-Gaussian) distribution. The binomial moments are given by

$$(n)_k B_k(n) = S(n, n-k)$$

Therefore, the mean $B_1(n)$ is $(n-1)/2$ while the variance is $(n+1)/12$.

## 7. SIMON NEWCOMB'S PROBLEM

As already described in Section 7.1, Simon Newcomb's problem is as follows: a deck of cards of arbitrary specification is dealt out into a single pile so long as the cards are in rising order, with like cards counted as rising, and a new pile is started whenever a non-rising card appears; with all possible arrangements of the deck, in how many ways do $k$ piles appear? Thus the problem is the same as the enumeration of permutations of $n$ numbered elements by number of ascending runs, when the elements are not necessarily distinct and like elements in sequence are counted as rising.

When all elements are distinct, the answer is given by the polynomials $A_n(t)$ of the preceding section, when the enumeration is by number of ascending runs, or by $A_{n,1}(t)$ when the enumeration is by number of descents. The board for the latter is the triangle of side $n-1$ with rook polynomial $T_{n-1}(x)$.

Enumeration by number of descents is more convenient for the general case, and will be followed below.

The board for the general case where the deck has specification $(1^{n_1}2^{n_2}\cdots s^{n_s})$ with $n_1 + 2n_2 + \cdots + sn_s = n$, that is, where there are $n_1$ distinct single elements, $n_2$ distinct pairs of like elements, $n_3$ triples and so on, may be obtained in the following way. Start with distinct elements 1 to $n$, specification $1^n$; the board specifying the descents is the triangle of side $n-1$. If two elements are taken as alike, and without loss of generality, they may be regarded as successive, as, for example, 1 and 2; then 21 is no longer a descent, and the cell at one end of the hypotenuse of the triangle is removed. Note that by Problem 7.4, all cells on the hypotenuse are equivalent, so it is irrelevant which pair of elements is identified. Similarly, if three elements, say 1, 2, 3, are identified, then 31, 32, and 21 are no longer descents, and a triangle of side two is removed. This triangle has two cells on the hypotenuse and one on its adjacent parallel, and, again, by Problem 7.4, the board is independent of the particular elements chosen for identification.

For an identification of $k+1$ successive elements, a triangle of side $k$ is removed, and the board for the general case is a triangle cut out to be a sort of ramp staircase. The boards for all specifications of 4 elements are as follows (dots indicate deleted cells):

```
×         ·         ·         ·         ·
× ×       × ×       × ·       · ·       · ·
× × ×     × × ×     × × ×     × × ×     · · ·
 1⁴        21²       2²        31        4
```

To determine the rook polynomials, suppose first that the specification is $1^{n-s}s$. Then the board may be taken as the triangle of side $n - 1$ with the first $s - 1$ rows removed, or, what is the same thing, as the trapezoid $(s, n - s, 1)$. The rook polynomial by equations (46) and (48) is then

$$\begin{aligned} R(1^{n-s}s, x) &= T(s, n - s, 1) \\ &= (T - x + sx)^{n-s} = T^{n-s-1}(T)_{s,x} \end{aligned} \tag{51}$$

with $(T)_{s,x} = T(T - x) \cdots (T - (s - 1)x)$, $T^n \equiv T_n(x)$, the triangular polynomial. The second form is more instructive because of its close relation to the specification and, indeed, suggests the general result:

$$R(1^{n_1}2^{n_2} \cdots s^{n_s}; x) = T^{n_1-1}(T)^{n_2}_{2,x} \cdots (T)^{n_s}_{s,x} \tag{52}$$

Before looking at the proof of this, notice that all specifications of 4 elements are covered by (51) except $2^2$, that the board for $2^2$ is a square with polynomial $1 + 4x + 2x^2$, and

$$T^{-1}(T)^2_2 = T(T - x)^2 = T_3 - 2xT_2 + x^2T_1 = 1 + 4x + 2x^2$$

in agreement with (52).

Indeed the board for specification $(1^{n-2k}2^k)$ is a triangle of side $n - 1$ with $k$ cells removed on the main diagonal, and

$$\begin{aligned} R(1^{n-2k}2^k, x) &= T_{n-1} - kxT_{n-2} + \cdots + \binom{k}{j}(-x)^jT_{n-1-j} + \cdots \\ &= T^{n-1-k}(T - x)^k, \qquad T^j \equiv T_j(x) \\ &= T^{n-2k-1}(T)^k_{2,x} \end{aligned}$$

in agreement with (52).

The proof of (51) is by induction in the following way. The specifications for any number of elements are ordered by decreasing number of parts of the corresponding partitions $(1^n)$, $(21^{n-2})$, $(31^{n-3})$, $(2^21^{n-4})$, and so on; so ordered, they are assembled according to rising values of $n$. The induction is carried on by expressing the polynomial for any specification in terms of those for some of its predecessors. To avoid complication of notation which adds nothing to the idea, the process is illustrated for some simple cases. The only specification of 5 elements not covered by proved results is (23). But by expansion indicated by

$$\begin{bmatrix} \times\times & & & \\ \times\times & \times & \times \end{bmatrix} = \begin{bmatrix} \times\times\times & \\ \times\times\times\times \end{bmatrix} - x\begin{bmatrix} \times\times\times \end{bmatrix}$$

$$\begin{aligned} R(23, x) &= R(1^23, x) - xR(13, x) \\ &= T(T)_3 - x(T)_3 = T(T-x)^2(T-2x) = T^{-1}(T)_2(T)_3 \end{aligned}$$

where, for brevity $(T)_k \equiv (T)_{k,x}$. Similarly

$$\begin{aligned} R(123, x) &= R(1^33, x) - xR(1^23, x) \\ &= T^2(T)_3 - xT(T)_3 = (T)_2(T)_3 \end{aligned}$$

Finally,

$$\begin{aligned} R(23^2, x) &= R(12^23, x) - 2xR(2^23, x) \\ &= (T)_2^2(T)_3 - 2xT^{-1}(T)_2^2(T)_3 \\ &= T^{-1}(T)_2(T)_3^2 \end{aligned}$$

It is clear from these examples that an expansion of the form required is implicit in the form of (51) and that there are numerous possibilities. For example, another form for the last example is

$$R(23^2, x) = R(1^23^2, x) - xR(13^2, x)$$

The hit polynomials, which supply the answer to Simon Newcomb's problem, as expected from (51), may be given a concise expression in terms of the triangular permutation polynomials $A_{n,1}(t)$ of the preceding section. Writing $[s]$ as an abbreviation for the specification $(1^{n_1}2^{n_2}\cdots s^{n_s})$, and $A_{[s]}(t)$ for the hit polynomial, this is

$$A_{[s]}(t) = 1!^{n_1}2!^{n_2}\cdots s!^{n_s}a_{[s]}(t) = A^{n_1}(A)_2^{n_2}\cdots(A)_s^{n_s} \tag{53}$$

with

$$(A)_s = A(A+1-t)(A+2-2t)\cdots(A+s-1-(s-1)t)$$

and

$$A^n \equiv A_n \equiv A_{n,1}(t)$$

The proof of (53), like that of (52), is by induction, and, again, to avoid complications of notation, illustrative examples are used. First for $[s] = (1^n)$, equation (53) is true by definition of $A_n(t)$. Next for $[s] = (21^{n-2})$

$$\begin{aligned} A_{[s]}(t) &= R(21^{n-2}, (t-1)/E)n! \\ &= T_{n-1}[(t-1)/E]n! - (t-1)T_{n-2}[(t-1)/E](n-1)! \\ &= A_n(t) + (1-t)A_{n-1}(t) \end{aligned}$$

Finally, to illustrate the induction

$$\begin{aligned} A_{[23^2]}(t) &= R(12^23, (t-1)/E)8! - 2(t-1)R(2^23, (t-1)/E)7! \\ &= A_{[12^23]}(t) + 2(1-t)A_{[2^23]}(t) \\ &= A(A)_2^2(A)_3 + 2(1-t)(A)_2^2(A)_3 = (A)_2(A)_3^2 \end{aligned}$$

Further illustrations of equation (53) appear in the problems.

## 8. THE PROBLEM OF THE BISHOPS

In how many ways can $k$ bishops be placed on an $n$ by $n$ chessboard so that no two attack each other? When it is remembered that bishops move on diagonals, it is evident that the board decomposes into two disjoint boards, the black and white squares, and that the enumerating polynomial for the whole board is the product of the polynomials for the black and white squares. Giving the board a 45° turn transforms the problem into one of rooks on a diamond-shaped board. Indicating black and white squares by $B$ and $W$ respectively, the two boards for $n = 3$ may be taken as

$$\begin{array}{cc} B & WW \\ BBB & WW \\ B & \end{array}$$

and for $n = 4$, they are

$$\begin{array}{cc} B & WW \\ BBB & WWWW \\ BBB & WW \\ B & \end{array}$$

which have the same rook polynomial.

Write $B_n(x)$ for the polynomial of the black squares, $W_n(x)$ for the white. Then rearranging the cells to form truncated triangles, it appears that

$$\begin{aligned} B_3(x) &= T(T)_2 = T^2(T-x) \\ W_3(x) &= T^{-1}(T)_2^2 = T(T-x)^2 \\ B_4(x) &= W_4(x) = (T)_2^2 = T^2(T-x)^2 \end{aligned}$$

which are all special cases of Simon Newcomb's problem.

The general results are

$$\begin{aligned} B_{2n}(x) &= W_{2n}(x) = (T)_2^n = T^n(T-x)^n, \qquad T^k \equiv T_k(x) \qquad (54) \\ B_{2n+1}(x) &= T(T)_2^n = T^{n+1}(T-x)^n \\ W_{2n+1}(x) &= T^{-1}(T)_2^{n+1} = T^n(T-x)^{n+1} \end{aligned}$$

By equation (42)

$$T^n(T-x) = (T+x)^n, \qquad T^k \equiv T_k(x)$$

Differentiation of (42) with respect to $y$ leads to

$$T^n(T - x)^2 = T(T + x)^n, \qquad T^k \equiv T_k(x)$$

which suggests that

$$T^n(T - x)^m = T^{m-1}(T + x)^n$$

a result readily verified by induction (using exponential generating functions).

With this result (54) may be given the alternative forms

$$\begin{aligned} B_{2n}(x) &= W_{2n}(x) = T^{n-1}(T + x)^n, \qquad T^k \equiv T_k(x) \qquad (54a) \\ B_{2n+1}(x) &= T^{n-1}(T + x)^{n+1} \\ W_{2n+1}(x) &= T^n(T + x)^n \end{aligned}$$

The first few instances of these are

$$\begin{aligned} B_2(x) &= 1 + 2x \\ B_3(x) &= 1 + 5x + 4x^2 \\ W_3(x) &= 1 + 4x + 2x^2 \\ B_4(x) &= 1 + 8x + 14x^2 + 4x^3 \\ B_5(x) &= 1 + 13x + 46x^2 + 46x^3 + 8x^4 \\ W_5(x) &= 1 + 12x + 38x^2 + 32x^3 + 4x^4 \\ B_6(x) &= 1 + 18x + 98x^2 + 184x^3 + 100x^4 + 8x^5 \\ B_8(x) &= 1 + 32x + 356x^2 + 1704x^3 + 3532x^4 + 2816x^5 + 632x^6 \\ &\quad + 16x^7 \end{aligned}$$

It may be noticed that

$$B_{2n+1}(x) = W_{2n+1}(x) + xB_{2n}(x)$$

## REFERENCES


1. L. Carlitz, Note on a paper of Shanks, *Amer. Math. Monthly*, vol. 59 (1952), pp. 239–241.
2. ———, Congruences for the ménage polynomials, *Duke Math. Journal*, vol. 19 (1952), pp. 549–552.
3. ———, Congruences connected with three-line Latin rectangles, *Proc. Amer. Math. Soc.*, vol. 4 (1953), pp. 9–11.
4. ———, Congruence properties of the ménage polynomials, *Scripta Mathematica*, vol. 20 (1954), pp. 51–57.
5. L. Carlitz and J. Riordan, Congruences for Eulerian numbers, *Duke Math. Journal*, vol. 20 (1953), pp. 339–344.
6. A. Cayley, On a problem of arrangements, *Proc. Royal Soc. Edinburgh*, vol. 9 (1878), pp. 338–342.

**7.** ———, Note on Mr. Muir's solution of a problem of arrangement, *Proc. Royal Soc. Edinburgh*, vol. 9 (1878), pp. 388–391.
**8.** P. Erdös and I. Kaplansky, The asymptotic number of Latin rectangles, *Amer. Journal of Math.*, vol. 68 (1946), pp. 230–236.
**9.** S. M. Jacob, The enumeration of the Latin rectangle of depth three, *Proc. London Math. Soc.*, vol. 31 (1930), pp. 329–354.
**10.** I. Kaplansky, Solution of the "Problème des ménages", *Bull. Amer. Math. Soc.*, vol. 49 (1943), pp. 784–785.
**11.** ———, Symbolic solution of certain problems in permutations, *Bull. Amer. Math. Soc.*, vol. 50 (1944), pp. 906–914.
**12.** I. Kaplansky and J. Riordan, The problème des ménages, *Scripta Mathematica*, vol. 12 (1946), pp. 113–124.
**13.** S. M. Kerawala, The enumeration of the Latin rectangle of depth three by means of a difference equation, *Bull. Calcutta Math. Soc.*, vol. 33 (1941), pp. 119–127.
**14.** ———, The asymptotic number of three-deep Latin rectangles, *Bull. Calcutta Math. Soc.*, vol. 39 (1947), pp. 71–72.
**15.** ———, Asymptotic solution of the "problème des ménages", *Bull. Calcutta Math. Soc.*, vol. 39 (1947), pp. 82–84.
**16.** E. Lucas, *Théorie des nombres*, Paris, 1891, pp. 491–495.
**17.** T. Muir, On Professor Tait's problem of arrangement, *Proc. Royal Soc. Edinburgh*, vol. 9 (1878), pp. 382–387.
**18.** ———, Additional note on a problem of arrangement, *Proc. Royal Soc. Edinburgh*, vol. 11 (1882), pp. 187–190.
**19.** N. S. Mendelsohn, The asymptotic series for a certain class of permutation problems, *Canadian Journal of Math.*, vol. 8 (1956), pp. 243–244.
**20.** E. Netto, *Lehrbuch der Combinatorik*, second edition, Berlin, 1927, pp. 75–80.
**21.** J. Riordan, Three-line Latin rectangles, *Amer. Math. Monthly*, vol. 51 (1944), pp. 450–452.
**22.** ———, Three-line Latin rectangles–II, *Amer. Math. Monthly*, vol. 53 (1946), pp. 18–20.
**23.** ———, Discordant permutations, *Scripta Mathematica*, vol. 20 (1954), pp. 14–23.
**24.** ———, Triangular permutation numbers, *Proc. Amer. Math. Soc.*, vol. 2 (1951), pp. 404–407.
**25.** ———, A recurrence relation for three-line Latin rectangles, *Amer. Math. Monthly*, vol. 59 (1952), pp. 159–162.
**26.** A. Sade, *Enumeration des carrés Latins de côté* 6, Marseille, 1948, 2 pp.
**27.** ———, *Enumeration des carrés Latins. Application au* 7$^{e}$ *ordre. Conjecture pour les ordres supérieurs*, Marseille, 1948, 8 pp.
**28.** ———, *Sur les suites hautes des permutations*, Marseille, 1949, 12 pp.
**29.** ———, *Décomposition des locomotions en facteurs de classe haute donnée*, Marseille, 1949, 8 pp.
**30.** ———, An omission in Norton's list of $7 \times 7$ squares, *Ann. Math. Statist.*, vol. 22 (1951), pp. 306–307.
**31.** L. v. Schrutka, Eine neue Einteilung der Permutationen, *Mathematische Annalen*, vol. 118 (1941), pp. 246–250.
**32.** W. Schöbe, Das Lucassche Ehepaarproblem, *Math. Zeitschrift*, vol. 48 (1943), pp. 781–784.
**33.** R. Sprague, Über ein Anordnungsproblem, *Mathematische Annalen*, vol. 121 (1949), pp. 52–53.
**34.** P. G. Tait, *Scientific Papers*, vol. 1, Cambridge, 1898, p. 287.

**35.** J. Touchard, Sur un problème de permutations, *C.R. Acad. Sci. Paris*, vol. 198 (1934), pp. 631–633.

**36.** ———, Permutations discordant with two given permutations, *Scripta Mathematica*, vol. 19 (1953), pp. 108–119.

**37.** J. Worpitzky, Studien über die Bernoullischen und Eulerschen Zahlen, *Journal für die reine und angewandte Math.*, vol. 94 (1883), pp. 203–232.

**38.** K. Yamamoto, An asymptotic series for the number of three-line Latin rectangles, *Journal Math. Soc. of Japan*, vol. 1 (1949), pp. 226–241.

**39.** ———, Latin Kukei no zenkinsu to symbolic method, *Sugaku*, vol. 2 (1944), pp. 159–162.

**40.** ———, Symbolic methods in the problem of three-line Latin rectangles, *Journal Math. Soc. of Japan*, vol. 5 (1953), pp. 13–23.

**41.** ———, On the asymptotic number of Latin rectangles, *Japanese Journal of Math.*, vol. 21 (1951), pp. 113–119.

**42.** ———, Structure polynomial of Latin rectangles and its application to a combinatorial problem, *Mem. Fac. Sci. Kyusyu Univ.*, ser. A, vol. 10 (1956), pp. 1–13.

## PROBLEMS

1. Generalizing the lemma in Section 2, show that the number of ways of selecting $k$ objects, no two "consecutive", from $n$ objects arrayed on a line, when $i + 1, \cdots, i + a$ are regarded as consecutive to $i$, is

$$f_a(n, k) = \binom{n - ak + a}{k}$$

and that the generating function

$$f_n(x; a) = \Sigma f_a(n, k)x^k$$

has the recurrence relation

$$f_n(x; a) = f_{n-1}(x; a) + xf_{n-a-1}(x; a)$$

2. For objects arrayed in a circle, show that the corresponding number $g_a(n, k)$ is given by

$$\begin{aligned} g_a(n, k) &= f_a(n - a, k) + af_a(n - 1 - 2a, k - 1) \\ &= \frac{n}{n - ak}\binom{n - ak}{k} = \frac{n}{k}\binom{n - ak - 1}{k - 1} \qquad \text{(Yamamoto, \textbf{42})} \end{aligned}$$

and that

$$g_n(x; a) = g_{n-1}(x; a) + xg_{n-a-1}(x; a)$$

3. (*a*) From equations (10) and (11) with $t = 0$ and $U_n(0) \equiv U_n$, show that

$$(n - 1)U_{n+1} = (n^2 - n + 1)(U_n + U_{n-1}) + nU_{n-2}$$

(*b*) Derive the same relation by iterating the corresponding instance of equation (19).

4. (*a*) Taking Chebyshev polynomials as defined by

$$T_n(x) = \cos n\theta, \qquad \cos\theta = x$$
$$U_n(x) = \sin (n + 1)\theta/\sin\theta, \qquad \cos\theta = x$$

with $T_0 = U_0 = 1$, show that

$$T_{n+1}(x) = 2xT_n(x) - T_{n-1}(x) = xT_n(x) - (1 - x^2)U_{n-1}(x)$$
$$U_{n+1}(x) = 2xU_n(x) - U_{n-1}(x) = xU_n(x) + T_{n+1}(x)$$

and verify the table

| $n$ | 0 | 1 | 2 | 3 | 4 | 5 |
|---|---|---|---|---|---|---|
| $T_n(x)$ | 1 | $x$ | $2x^2 - 1$ | $4x^3 - 3x$ | $8x^4 - 8x^2 + 1$ | $16x^5 - 20x^3 + 5x$ |
| $U_n(x)$ | 1 | $2x$ | $4x^2 - 1$ | $8x^3 - 4x$ | $16x^4 - 12x^2 + 1$ | $32x^5 - 32x^3 + 6x$ |

(*b*) Derive the reciprocal relations

$$t_n(x) = \sum_0^m (-1)^k \frac{n}{n-k}\binom{n-k}{k}(2x)^{n-2k}, \qquad m = [n/2]$$

$$(2x)^n = \sum_0^m \binom{n}{k} t_{n-2k}(x), \qquad m = [n/2]$$

with $t_0(x) = T_0$, $t_n(x) = 2T_n(x)$, $n > 0$.
and

$$U_n(x) = \sum_0^m (-1)^k \binom{n-k}{k}(2x)^{n-2k}$$

$$(2x)^n = \sum_0^m \left[\binom{n}{k} - \binom{n}{k-1}\right] U_{n-2k}(x)$$

(*c*) Derive the generating functions

$$\begin{aligned} T(x, y) &= 1 + xy + T_2(x)y^2 + \cdots + T_n(x)y^n + \cdots \\ &= (1 - xy)(1 - 2xy + y^2)^{-1} \\ U(x, y) &= 1 + 2xy + U_2(x)y^2 + \cdots + U_n(x)y^n + \cdots \\ &= (1 - 2xy + y^2)^{-1} \end{aligned}$$

5. (*a*) Using equations (8) and (9), show that

$$\begin{aligned} L(x, y) &= 1 + L_1(x)y + L_3(x)y^2 + \cdots + L_{2n-1}(x)y^n + \cdots \\ &= (1 - xy)(1 - y - 2xy + x^2y^2)^{-1} \\ M(x, y) &= 2 + (1 + 2x)y + M_2(x)y^2 + \cdots + M_n(x)y^n + \cdots \\ &= (2 - y - 2xy)(1 - y - 2xy + x^2y^2)^{-1} \\ &= 1 + (1 - x^2y^2)(1 - y - 2xy + x^2y^2)^{-1} \\ &= 1 + (1 + xy)L(x, y) \end{aligned}$$

and

$$l(x, y) = L(-x^{-1}, xy) = (1 + y)(1 - xy + 2y + y^2)^{-1}$$
$$m(x, y) = M(-x^{-1}, xy) = (2 - xy + 2y)(1 - xy + 2y + y^2)^{-1}$$

(*b*) Expanding $(1 - y - 2xy + x^2y^2)^{-1}$ in partial fractions, find the expressions

$$L_{2n-1}(x) = \frac{(1 + \alpha)^{2n+1} - (1 - \alpha)^{2n+1}}{2^{2n+1}\alpha}, \qquad \alpha = \sqrt{1 + 4x}$$

$$M_n(x) = \left(\frac{1 + 2x + \alpha}{2}\right)^n + \left(\frac{1 + 2x - \alpha}{2}\right)^n$$

(c) With $xy = t$, $(u - 1)2x = 1$, and Chebyshev polynomial notation as in Problem 4, show that

$$L(u, t) = (1 - t)(1 - 2tu + t^2)^{-1} = (1 - t)U(u, t)$$
$$M(u, t) = 2(1 - tu)(1 - 2tu + t^2)^{-1} = 2T(u, t)$$

so that

$$L_{2n-1}(x) = x^n \left[U_n\left(1 + \frac{1}{2x}\right) - U_{n-1}\left(1 + \frac{1}{2x}\right)\right]$$

$$M_n(x) = 2x^n T_n\left(1 + \frac{1}{2x}\right)$$

6. Define the even Chebyshev polynomial generating functions by

$$2T_e(x, y) = T(x, y) + T(x, -y) = 2\Sigma T_{2n}(x)y^n$$
$$2U_e(x, y) = U(x, y) + U(x, -y) = 2\Sigma U_{2n}(x)y^n$$

Show that

$$[1 + 2y^2 - 4x^2y^2 + y^4]T_e(x, y) = 1 + y^2 - 2x^2y^2$$
$$[1 + 2y^2 - 4x^2y^2 + y^4]U_e(x, y) = 1 + y^2$$

and that in the notation of Problem 5,

$$L(-x^{-1}, xy) = U_e(\sqrt{x}/2, \sqrt{y})$$
$$M(-x^{-1}, xy) = 2T_e(\sqrt{x}/2, \sqrt{y})$$

Hence

$$x^n L_{2n-1}(-1/x) = U_{2n}(\sqrt{x}/2)$$
$$x^n M_n(-1/x) = 2T_{2n}(\sqrt{x}/2)$$

7. (a) Using the result of Problem 5(c), show that the ménage hit polynomial is given by

$$U_n(t) = 2(t - 1)^n T_n\left(\frac{2t + E - 2}{2t - 2}\right) 0!, \qquad E^k 0! = k!$$

$$= 2(t - 1)^n T_n\left(\frac{2t + e}{2t - 2}\right), \qquad e^n \equiv e_n = (E - 2)^n 0!$$

(compare Problem 7.8 for the numbers $e_n$), so that $U_1(t) = 2t + e_1 = -1 + 2t$, $U_2(t) = (2t + e)^2 - 2(t - 1)^2 = 2t^2$.

In particular

$$U_n(0) \equiv U_n = 2T_n(e/2), \qquad e^n \equiv e_n$$

$$= \sum_0^m (-1)^k \frac{n}{n - k}\binom{n - k}{k} e_{n-2k}, \qquad m = [n/2]$$

(Yamamoto, **42**)

if $U_0 = e_0 = 1$, and (compare Problem 4(b))

$$e_n = \sum_0^m \binom{n}{k} U_{n-2k}, \qquad U_0 = 1, \qquad m = [n/2]$$

or

$$2e_n = (U + U^{-1})^n, \qquad U_0 = 2, \qquad U^n = U^{-n} \equiv U_n$$

(*b*) Derive the following asymptotic expression

$$U_n \sim n!\, e^{-2}\left(1 - \frac{1}{n-1} + \frac{1}{2!\,(n-1)_2} + \cdots + \frac{(-1)^k}{k!\,(n-1)_k} + \cdots\right)$$

with $(n)_k = n(n-1)\cdots(n-k+1)$.

8. (*a*) Write, as in equation (22),

$$l_n(x) = x^n L_{2n-1}(-x^{-1})$$
$$m_n(x) = x^n M_n(-x^{-1})$$

Show that

$$x^n = \sum_0^n \left[\binom{2n}{k} - \binom{2n}{k-1}\right] l_{n-k}(x)$$
$$= \sum_0^n \binom{2n}{k} m_{n-k}(x), \qquad m_0 = 1$$

(*b*) Derive the inverse formulas

$$n! = \sum_0^n \left[\binom{2n}{k} - \binom{2n}{k-1}\right] (1-t)^k V_{n-k}(t)$$
$$= \sum_0^n \binom{2n}{k} (1-t)^k U_{n-k}(t)$$

where $U_n(t)$ and $V_n(t)$ are ménage hit polynomials and $U_0(t) = 1$.

9. (*a*) From the recurrence, equation (13),

$$(n-1)kU_{n,k} = n(2n-1-k)U_{n-1,k-1} + nkU_{n-1,k}$$

derive the results

$$U_{n,n} = 2 \qquad U_{n,n-2} = 4\binom{n}{4}$$

$$U_{n,n-1} = n(n-2) \qquad U_{n,n-3} = \binom{n}{3} + 4\binom{n}{4} + 10\binom{n}{5} + 12\binom{n}{6}$$

(*b*) Similarly, from

$$kV_{n,k} = kV_{n-1,k} + (2n-k)V_{n-1,k-1}$$

derive

$$V_{n,n} = 1 \qquad V_{n,n-2} = \binom{n}{3} + 2\binom{n}{4}$$

$$V_{n,n-1} = \binom{n}{2} \qquad V_{n,n-3} = \binom{n}{3} + 4\binom{n}{4} + 8\binom{n}{5} + 6\binom{n}{6}$$

10. (*a*) From recurrences (8) and (9), show that the coefficients $A_{kj}$, $B_{kj}$ in expressions

$$L_{2n-1,k} = \sum_{j=0} A_{kj}\binom{n-j}{k-j} = \binom{2n-k}{k}$$

$$M_{nk} = \sum_{j=0} B_{kj}\binom{n-j}{k-j} = \frac{2n}{2n-k}\binom{2n-k}{k}$$

satisfy recurrences

$$A_{kj} = 2A_{k-1,j} - A_{k-2,j-1}$$
$$B_{kj} = 2B_{k-1,j} - B_{k-2,j-1}$$

Verify the initial values

$$L_{2n-1,0} = 1 \qquad M_{n0} = 1$$

$$L_{2n-1,1} = 2n - 1 \qquad M_{n1} = 2n$$

$$L_{2n-1,2} = 4\binom{n}{2} - 3(n-1) \qquad M_{n2} = 4\binom{n}{2} - (n-1) - 1$$

$$L_{2n-1,3} = 8\binom{n}{3} - 8\binom{n-1}{2} + (n-2) \qquad M_{n3} = 8\binom{n}{3} - 4\binom{n-1}{2} - 2(n-2)$$

(*b*) Show that

$$B_{kj} = A_{kj} + A_{k-1,j-1}$$

(*c*) Derive the generating function relations

$$a_k(x) = \Sigma 2^{-k} A_{kj} x^j = a_{k-1}(x) - (x/4)a_{k-2}(x)$$
$$b_k(x) = \Sigma 2^{-k} B_{kj} x^j = b_{k-1}(x) - (x/4)b_{k-2}(x)$$

and, hence, show that

$$a_0 = b_0 = 1$$
$$a_k(x) = L_{k-1}(-x/4) - (x/2)L_{k-2}(-x/4)$$
$$= L_k(-x/4) - (x/4)L_{k-2}(-x/4)$$
$$b_k(x) = L_{k-1}(-x/4) - (x^2/4)L_{k-3}(-x/4)$$

with $L_k(x)$ the staircase polynomial defined by (1).

(*d*) Show that

$$A_{kj} = (-1)^j 2^{k-2j}\left[\binom{k-j}{j} + 2\binom{k-j}{j-1}\right], \qquad j \leq [k/2]$$

$$B_{kj} = (-1)^j 2^{k-2j}\left[\binom{k-j}{j} - 4\binom{k-j}{j-2}\right], \qquad j \leq [k/2]$$

$$\Sigma A_{kj} = 1$$
$$\Sigma B_{kj} = 2$$

11. Using the expression for probabilities in terms of factorial moments, equation (2.32), and the relation

$$r_k = \binom{n}{k} M_{(k)}$$

with $r_k$ a rook polynomial coefficient and $M_{(k)}$ the $k$th factorial moment of the hit distribution, show that, in the notation of Problem (10), the factorial moments for straight and circular tables are given by

$$M_{(k)} = \sum_{j=0} A_{kj}(k)_j/(n)_j \qquad \text{(straight table)}$$

$$= \sum_{j=0} B_{kj}(k)_j/(n)_j \qquad \text{(circular table)}$$

Derive the series

$$\frac{V_{nr}}{n!} = \frac{2^r e^{-2}}{r!} \sum_{j=0} \frac{(-1)^j}{(n)_j} v_j(r)$$

$$\frac{U_{nr}}{n!} = \frac{2^r e^{-2}}{r!} \sum_{j=0} \frac{(-1)^j}{(n-1)_j} F_j(r) = \frac{2^r e^{-2}}{r!} \sum_{j=0} \frac{(-1)^j}{(n)_j} u_j(r)$$

with

$$4^j j!\, F_j(r) = [(r) - 2]^{2j}, \qquad (r)^j \equiv (r)_j$$

$$4^j j!\, u_j(r) = [(r) - 2]^{2j} - 4j(j-1)[(r) - 2]^{2j-2}$$

$$4^j j!\, v_j(r) = [(r) - 2]^{2j} + 2j[(r) - 2]^{2j-1}, \qquad (r)^j \equiv (r)_j$$

12. Show that the staircase polynomials $L_k(x)$ defined by $L_0 = 1$, $L_1 = 1 + x$, and [equation (1)]

$$L_k(x) = L_{k-1}(x) + xL_{k-2}(x)$$

satisfy

$$L_{j+k+1}(x) = L_j L_k + xL_{j-1}L_{k-1}$$

and

$$L_{j+k} = L_j L_k - x^2 L_{j-2} L_{k-2}$$

13. Write $f_n(r) = [(r) - 2]^n$, $(r)^n \equiv (r)_n$; show that

$$\exp xf(r) = \sum_0^\infty f_n(r) \frac{x^n}{n!} = e^{-2x}(1 + x)^r$$

and derive the recurrences

$$f_{n+1} = (r - n - 2)f_n - 2nf_{n-1}$$
$$= -2f_n + rg_n$$

with $g_n = [(r - 1) - 2]^n$. Verify the initial values

| | |
|---|---|
| $f_0 = 1$ | $g_0 = 1$ |
| $f_1 = r - 2$ | $g_1 = r - 3$ |
| $f_2 = r^2 - 5r + 4$ | $g_3 = r^2 - 7r + 10$ |
| $f_3 = r^3 - 9r^2 + 20r - 8$ | $g_4 = r^3 - 12r^2 + 41r - 38$ |

14. Using the relation, equation (7),

$$U_n(t) = V_n(t) - (1 - t)V_{n-1}(t)$$

and the notation of Problem (11), show that

$$u_j(r) = v_j(r) + v_{j-1}(r) - (r/2)v_{j-1}(r - 1)$$

Use the second recurrence of Problem 13 to verify the results of Problem 11.

15. For the polynomials $m_n(x)$ of Problem 8, show that

$$m_n(x) = m_1(x)m_{n-1}(x) - m_{n-2}(x)$$
$$= m_2(x)m_{n-2}(x) - m_{n-4}(x)$$

and, by induction,

$$m_n(x) = m_k(x)m_{n-k}(x) - m_{n-2k}(x)$$

in agreement with equation (25).

16. Show that

$$[m_n(x)[^{2k} = \sum_0^{k-1} \binom{2k}{j} m_{2n(k-j)}(x) + \binom{2k}{k}$$

$$[m_n(x)]^{2k+1} = \sum_0^{k} \binom{2k+1}{j} m_{n+2n(k-j)}(x)$$

17. The board for permutations discordant with

$$\begin{matrix} 1 & 2 & \cdots & n-1 & n \\ n & n-1 & \cdots & 2 & 1 \end{matrix}$$

consists of the two main diagonals of a square of side $n$, and has rook polynomial $[M_2(x)]^p$ for $n = 2p$, $(1 + x)[M_2(x)]^p$ for $n = 2p + 1$. With $A_n(t)$ the corresponding hit polynomial as in Problems 7.7 and 7.9, show that

$$A_{2p}(t) = \sum_{j=0}^{q} \binom{p}{j} (1 - t)^{4j} U_{2p-4j}(t), \qquad q = [p/2]$$

$$A_{2p+1}(t) = \sum_{j=0}^{q} \binom{p}{j} (1 - t)^{4j} u_{2p+1-4j}(t), \qquad q = [p/2]$$

with $U_n(t)$ the ménage hit polynomial, $U_0 = 1$, and

$$u_n(t) = U_n(t) + (1 - t)U_{n-1}(t) + (1 - t)^2 U_{n-2}(t)$$

Verify the instances

$$A_1(t) = t \qquad A_3(t) = 2(2t + t^3)$$

$$A_2(t) = 2t^2 \qquad A_4(t) = 4(1 + 4t^2 + t^4)$$

and compare with Problems 7.7 and 7.9.

18. Using the results of Problem 6 and Euler's integral

$$(n - k)! = \int_0^\infty e^{-x} x^{n-k}\, dx$$

show that

$$U_n = \int_0^\infty e^{-x} m_n(x)\, dx = 16 \int_0^\infty e^{-4x^2} x T_{2n}(x)\, dx$$

$$V_n = \int_0^\infty e^{-x} l_n(x)\, dx = 8 \int_0^\infty e^{-4x^2} x U_{2n}(x)\, dx$$

[$U_n$, $V_n$ are circular- and straight-table ménage numbers; $m_n(x)$ and $l_n(x)$ are associated rook polynomials as in Problem 6; $T_n(x)$ and $U_n(x)$ are Chebyshev polynomials.]

19. Write the relation

$$n! = \sum_0^n \binom{2n}{k} (1 - t)^k U_{n-k}(t)$$

of Problem 8(*b*) as

$$n! = (U^{1/2} + (1 - t)U^{-1/2})^{2n}, \qquad U^n \equiv U_n(t), \qquad U_{-n} \equiv 0, \qquad n > 0$$

Using the generating function for Bessel functions of the first kind with imaginary argument, namely,

$$\exp -t(u + u^{-1}) = \sum_{-\infty}^{\infty} u^n I_n(2t)$$

show that

$$(1 - x)^{-1} \exp -2x(1 - t) = \sum_{0}^{\infty} (1 - t)^{-n} U_n(t) I_n(2x - 2tx)$$

20. (*a*) Write $P_n(x)$ for the rook polynomial of a board containing the $2(n - 1)$ cells adjacent to the main diagonal of a square of side $n$; for example, the boards for $n = 2$, 3, and 4 are

```
                               . ×
  . ×        . ×             × . ×
  × .        × . ×             × . ×
               × .               × .
```

Write $P_n^*(x)$ for the polynomial of the same board without the cell in the first column. With $L_n(x)$ the staircase polynomial of equation (1), show that for $n > 2$

$$P_n(x) = (1 + x)P_n^*(x) - x^2 L_{n-1}(x) L_{n-4}(x)$$
$$P_n^*(x) = P_{n-1}(x) + xP_{n-1}^*(x)$$

so that

$$P_n(x) = (1 + 2x)P_{n-1}(x) + x^3 L_{n-2} L_{n-5} - x^2 L_{n-1} L_{n-4}$$

Note that, by inspection, $P_2(x) = (1 + x)^2$, $P_2^*(x) = 1 + x$, $P_3(x) = (1 + 2x)^2$, $P_3^*(x) = (1 + x)(1 + 2x)$; hence, the first equation holds for $n > 1$ if $L_{-2}(x) = 0$, $L_{-1}(x) = 1$, and, for the same range, the seconds holds if $P_1(x) = P_1^*(x) = 1$, a natural convention.

(*b*) Using Problem 12, derive in succession

$$P_n(x) = (1 + 2x)P_{n-1}(x) - x^2(L_{2n-4}(x) - 2xL_{2n-6}(x) + \cdots + 2(-x)^j L_{2n-4-2j}(x) + \cdots)$$

$$P_n(x) + xP_{n-1}(x) = (1 + 2x)(P_{n-1}(x) + xP_{n-2}(x)) - x^2 L_{2n-5}(x) = L_{2n-1}(x)$$

$$P_n(x) = (1 + x)P_{n-1}(x) + x(1 + x)P_{n-2}(x) - x^3 P_{n-3}(x)$$

(*c*) With $P_0(x) = P_1(x) = 1$, show that

$$P(x, y) = P_0(x) + P_1(x)y + P_2(x)y^2 + \cdots$$
$$= L(x, y)(1 + xy)^{-1} = (1 - xy)(1 + xy)^{-1}(1 - y - 2xy + x^2y^2)^{-1}$$

and that, in the notation of Problem 5

$$p(x, y) = P(-x^{-1}, xy) = \Sigma x^n P_n(-1/x)y^n = \sum_{0}^{\infty} p_n(x)y^n$$
$$= l(x, y)(1 - y)^{-1} = (1 + y)(1 - y)^{-1}(1 + 2y + y^2 - xy)^{-1}$$

(*d*) From the last relation, derive the relation

$$p_n(x) - p_{n-1}(x) = l_n(x)$$

and, hence,

$$A_n(t) = (1 - t)^n p_n(E(1 - t)^{-1})0!$$
$$= V_n(t) + (1 - t)A_{n-1}(t)$$

with $V_n(t)$ the hit polynomial for the straight-table ménages problem. Verify the initial values:

$$A_0 = 1 \qquad A_2 = 1 + t^2$$
$$A_1 = 1 \qquad A_3 = 2 + 4t^2$$

(*e*) Using equation (20) and the result in (*d*), show that

$$\begin{aligned}(n-1)A_n(t) = {} & [n^2 - 2 - (n-2)t]A_{n-1}(t) \\ & - [n^2 - 2n - 1 + (n+1)t](1-t)A_{n-2}(t) \\ & - n(1-t)^3A_{n-3}(t) - 2(t-1)^n, \qquad n > 1\end{aligned}$$

(*f*) Write

$$\frac{A_{n,r}}{n!} = \frac{2^r e^{-2}}{r!} \sum_{j=0} \frac{(-1)^j}{(n)_j} a_j(r)$$

Using the relation of (*d*) above and Problem 14, show that

$$v_j(r) = a_j(r) + a_{j-1}(r) - (r/2)a_{j-1}(r-1)$$

where $4^j j!\, v_j(r) = f_{2j} + 2jf_{2j-1}$, with $f_n$ as in Problem 13. Derive the results

$$\begin{aligned}a_0(r) &= 1 \\ 4a_1(r) &= f_2 + 4f_1 \\ 32a_2(r) &= f_4 + 8f_3 + 16f_2 \\ 4^j j!\, a_j(r) &= f_{2j} + 4jf_{2j-1} + 8(j)_2 f_{2j-2} + \cdots + 2^{k+1}(j)_k f_{2j-k} + \cdots 2^{j+1} j!\, f_j\end{aligned}$$

21. Write $S_k(x)$ for the rook polynomial of a three-ply staircase

$$\begin{matrix}\times\times\times & & & \\ & \times\times\times & & \\ & & \cdots & \\ & & & \times\times\times \\ & & & & \times\times\times\end{matrix}$$

with $k$ rows and $k+2$ columns. Note that $S_1(x) = 1 + 3x$, $S_2(x) = 1 + 6x + 7x^2$. Expanding with respect to cells in the first row, show that

$$S_k(x) = (1+x)S_{k-1}(x) + x(S^*_{k-1}(x) + S^{**}_{k-1}(x))$$

with $S^*_{k-1}(x)$ the polynomial of the board with first row and second column removed, $S^{**}_{k-1}(x)$ the polynomial of the board with first row and third column removed. Show further that

$$\begin{aligned}S_k(x) &= S^*_k(x) + xS_{k-1}(x) \\ &= S^{**}_k(x) + x(S^*_{k-1}(x) + S_{k-1}(x)) - x^2S^{**}_{k-2}(x)\end{aligned}$$

Hence, by elimination,

$$S_k(x) = (1+3x)S_{k-1}(x) - 2x^2S_{k-2}(x) - x^2(1+x)S_{k-3}(x) + x^4S_{k-4}(x)$$

Verify that this holds for all $k$ if $S_0(x) = 1$, $S_{-n}(x) = 0$, $n > 0$.

22. Write $T_k(x)$ for the polynomial of the staircase in Problem 21 with the first and last column removed. Show that

$$T_k(x) = S_k(x) - 2xS_{k-1}(x) + x^2S_{k-2}(x)$$

so that $T_0 = 1$, $T_1 = 1 + x$, $T_2 = 1 + 4x + 2x^2$. Hence,

$$T_k(x) = (1+3x)T_{k-1}(x) - 2x^2T_{k-2}(x) - x^2(1+x)T_{k-3}(x) + x^4T_{k-4}(x)$$

23. (*a*) Removing the main diagonal of the board above for $T_k(x)$ produces the board for $P_k(x)$ of Problem 20. Expanding with respect to the cells on the main diagonal, show that

$$T_n(x) = P_n(x) + x\sum_{k=0} T_k(x)P_{n-1-k}(x)$$

$$\begin{aligned} T(x, y) &= \sum_{n=0}^{\infty} T_n(x)y^n \\ &= P(x, y) + xyP(x, y)T(x, y) \\ &= P(x, y)[1 - xyP(x, y)]^{-1} \\ &= (1 - xy)[1 - y - 2xy - xy^2 + x^3y^3]^{-1} \end{aligned}$$

using Problem 20(*c*) for the last line.

(*b*) Derive the last expression from the recurrence relation of Problem 22, and verify that, in the notation of Problem 21,

$$\begin{aligned} S(x, y) &= \Sigma S_n(x)y^n \\ &= T(x, y)(1 - xy)^{-2} \\ &= (1 - xy)^{-1}(1 - y - 2xy - xy^2 + x^3y^3)^{-1} \end{aligned}$$

24. The permutations discordant with the three permutations

$$\begin{matrix} 1 & 2 & \cdots & n-1 & n \\ 2 & 3 & \cdots & n & 1 \\ 3 & 4 & \cdots & 1 & 2 \end{matrix}$$

have a board which is the three-ply staircase of Problem 21 with the triangle in the last two columns removed to the first two columns; for example, the board for $n = 5$ is

```
× × ×
  × × ×
    × × ×
×     × ×
× ×     ×
```

Write $R_n(x)$ for its rook polynomial.

(*a*) If $s_n(x)$ is the polynomial for the three-ply staircase [polynomial $S_n(x)$] with the last two columns deleted, show that

$$s_n(x) = S_{n-1}(x) - xS_{n-2}(x) + xs_{n-1}(x)$$

so

$$s_n(x) - S_{n-1}(x) = x(s_{n-1}(x) - S_{n-2}(x)) = x^n$$

giving

$$s_0 = 1 \qquad s_2 = 1 + 3x + x^2$$

$$s_1 = 1 + x \qquad s_3 = 1 + 6x + 7x^2 + x^3$$

From this, show that

$$\begin{aligned} s(x, y) &= \sum_{n=0} s_n(x)y^n \\ &= yS(x, y) + (1 - xy)^{-1} \\ &= (1 - 2xy - xy^2 + x^3y^3)(1 - xy)^{-1}(1 - y - 2xy - xy^2 + x^3y^3)^{-1} \\ &= (1 - 2xy - xy^2 + x^3y^3)S(x, y) \end{aligned}$$

and

$$s_n(x) = S_n(x) - 2xS_{n-1}(x) - xS_{n-2}(x) + x^3S_{n-3}(x)$$

(*b*) Developing the board for $R_n(x)$ with respect to the triangle in the lower left-hand corner, show that

$$R_n(x) = s_n(x) + xT_{n-1}(x) + 2x\tau_{n-1}(x) + x^2s_{n-2}(x)$$

[$T_n(x)$ is the polynomial of Problem 22]. $\tau_{n-1}(x)$ is the polynomial for the board with the first column and next to last row deleted,

$$\tau_n(x) = S_{n-1}(x) - xS_{n-2}(x) + xT_{n-1}(x)$$

so

$$\begin{aligned} R_n(x) &= S_n(x) - xS_{n-1}(x) + (x + x^2)S_{n-2}(x) - 2x^2(1 + 2x)S_{n-3}(x) \\ &\quad - x^3(1 - 2x)S_{n-4}(x) + x^5S_{n-5}(x) \\ &= S_n(x) - 2x^2S_{n-2}(x) - 2x^2(1 + x)S_{n-3}(x) + 3x^4S_{n-4}(x) \end{aligned}$$

and, if $R_0 = S_0 = 1$, $R_1 = 1 + 3x$, $R_2 = 1 + 6x + 5x^2$,

$$\begin{aligned} R(x, y) &= \Sigma R_n(x)y^n \\ &= [1 - 2x^2y^2 - 2x^2(1 + x)y^3 + 3x^4y^4]S(x, y) \end{aligned}$$

and

$$\begin{aligned} r(x, y) &= R(-x^{-1}, xy) \\ &= (1 - 2y^2 - 2(x - 1)y^3 + 3y^4)/s(x, y) \end{aligned}$$

with

$$\begin{aligned} s(x, y) &= S^{-1}(-x^{-1}, xy) \\ &= (1 + y)[(1 + y)(1 + y - y^2) - xy(1 - y)] \end{aligned}$$

Note that this is *not* the $s(x, y)$ of part (*a*). Note also that $r_0(x) = 1$, $r_1(x) = x - 3$, $r_2(x) = x^2 - 6x + 5$.

25. Continuing the problem above, show that, with $r_0(x) = 4$, and a new definition of $r(x, y)$, namely,

$$\begin{aligned} r(x, y) &= 4 + r_1(x)y + r_2(x)y^2 + \cdots + r_n(x)y^n + \cdots \\ &= (4 - 3(x - 3)y + 4y^2 + (x - 1)y^3)/s(x, y) \\ &= 4 - ys_y(x, y)/s(x, y) \end{aligned}$$

with $s_y(x, y)$ a partial derivative as usual. Show further that (by partial fraction expansion)

$$\begin{aligned} r^*(x, y) &= r(x, y) - (1 + y)^{-1} \\ &= [3 + 4y - xy(2 - y)]/\Delta(x, y) \\ &= 3 - y\Delta_y(x, y)/\Delta(x, y) \end{aligned}$$

with $\Delta(x, y) = s(x, y)(1 + y)^{-1} = (1 + y)(1 + y - y^2) - xy(1 - y)$, and, if

$$r_n'(x) = nq_{n-1}(x)$$

with the prime denoting a derivative,

$$\begin{aligned} q(x, y) &= q_0(x) + q_1(x)y + \cdots + q_n(x)y^n + \cdots \\ &= (1 - y)/\Delta(x, y) \end{aligned}$$

Derive the recurrence relations

$$r_n(x) - (x - 2)r_{n-1}(x) + xr_{n-2}(x) - r_{n-3}(x) = 2x(-1)^n, \qquad n > 2$$
$$q_n(x) - (x - 2)q_{n-1}(x) + xq_{n-2}(x) - q_{n-3}(x) = 0, \qquad n > 1$$

26. (*a*) Continuing, show that the generating functions above satisfy

$$(1 - y)^2 r^*(x, y) - (1 - 2y - 2y^2 + 2y^3 - y^4)q(x, y) = 2 - 3y + y^2$$

Derive the recurrence

$$r_n(x) - 2r_{n-1}(x) + r_{n-2}(x) - 4(-1)^n = q_n(x) - 2q_{n-1}(x) - 2q_{n-2}(x)$$
$$+ 2q_{n-3}(x) - q_{n-4}(x), \qquad n > 2$$

(Yamamoto, **42**)

(*b*) Show that

$$r(0, y) = \frac{4 + 5y - y^2}{(1 + y)(1 + y - y^2)} = \frac{2}{1 + y} + \frac{2 + y}{1 + y - y^2}$$

$$q(0, y) = \frac{1 - y}{(1 + y)(1 + y - y^2)} = \frac{-2}{1 + y} + \frac{3 - 2y}{1 + y - y^2}$$

Hence, if $f_n$ is a Fibonacci number ($f_n = f_{n-1} + f_{n-2}$) with $f_0 = f_1 = 1$,

$$r_n(0) = (-1)^n(2 + 2f_n - f_{n-1})$$
$$q_n(0) = (-1)^n(-2 + 3f_n + 2f_{n-1})$$

Write $(-1)^n r_n(0) = a_n$, $(-1)^n q_n(0) = b_n$; derive the recurrences

$$a_n = a_{n-1} + a_{n-2} - 2, \qquad n > 1$$
$$b_n = b_{n-1} + b_{n-2} + 2, \qquad n > 1$$

and verify the table

| $n$ | 0 | 1 | 2 | 3 | 4 | 5 | 6 | 7 | 8 | 9 | 10 |
|---|---|---|---|---|---|---|---|---|---|---|---|
| $a_n$ | 4 | 3 | 5 | 6 | 9 | 13 | 20 | 31 | 49 | 78 | 125 |
| $b_n$ | 1 | 3 | 6 | 11 | 19 | 32 | 53 | 87 | 142 | 231 | 375 |

27. For the hit polynomials corresponding to $r_n(x)$ and $q_n(x)$ of Problem 26, write

$$N_n(t) = (1 - t)^n r_n[E(1 - t)^{-1}]0! = \Sigma N_{nr} t^r$$
$$M_n(t) = (1 - t)^n q_n[E(1 - t)^{-1}]0! = \Sigma M_{nr} t^r$$

Using $r_n'(x) = nq_{n-1}(x)$ and equation (7.4), derive

$$(n + 1)M_n(t) = N_{n+1}(t) - (1 - t)^{n+1} r_{n+1}(0)$$

and verify the table

| $n$ | 0 | 1 | 2 | 3 | 4 |
|---|---|---|---|---|---|
| $N_n(t)$ | 4 | $-2+3t$ | $1-4t+5t^2$ | $6t^3$ | $1+ \quad 6t^2+ 8t^3+ 9t^4$ |
| $M_n(t)$ | 1 | $-2+3t$ | $2-6t+6t^2$ | $-2+9t-12t^2+11t^3$ | $3-10t+30t^2-18t^3+19t^4$ |

Using the recurrence of Problem 26(*a*), derive the relation

$$N_n(t) - 2(1 - t)N_{n-1}(t) + (1 - t)^2 N_{n-2}(t) - 4(t - 1)^n$$
$$= M_n(t) - 2(1 - t)M_{n-1}(t) - 2(1 - t)^2 M_{n-2}(t)$$
$$+ 2(1 - t)^3 M_{n-3}(t) - (1 - t)^4 M_{n-4}(t)$$

(Yamamoto, **42**)

28. Continuing, derive the asymptotic expression

$$\frac{N_{nr}}{n!} = \frac{3^r e^{-3}}{r!}\left(1 - \frac{r^2 - 7r + 9}{3n} + \frac{a_2(r)}{18n(n-1)}\right) + 0(n^{-3})$$

with

$$a_2(r) = 24\binom{r}{4} - 64\binom{r}{3} + 84\binom{r}{2} - 78r - 27$$

29. The complement of a triangle of side $n$ in an $n$ by $n$ square is a triangle of side $n - 1$. Hence with $T_n(x)$ the triangular rook polynomial, $S_n(x)$ the square polynomial, and $S(n, k)$ the Stirling number of the second kind

$$T_n(x) = \sum_{k=0}^{n} S(n, n-k)(-x)^k S_{n-k}(x)$$

$$T_{n-1}(x) = \Sigma S(n+1, n+1-k)(-x)^k S_{n-k}(x)$$

Show by inversion of these equations that

$$S_n(x) = T(T+x)\cdots(T+(n-1)x), \qquad T^k \equiv T_k(x)$$

$$= T^{-1}(T+x)\cdots(T+nx), \qquad T^k \equiv T_k(x), \qquad T_{-1} = 1$$

30. Write $U_k(p, q, 1; x)$ for the polynomial of the truncated trapezoid

```
      ←   p   →
  ↑  × × · · · ×
     × × · · · × ×
  q  · · · · · · · · ·
     × × · · · × × · · · ×   ↑
     × × · · · × × · · · ×
     · · · · · · · · · · · ·  k
  ↓  × × · · · × × · · · ×   ↓
      ←    p + q − k    →
```

which is the complement of the triangle of side $q - k$ in the rectangle $q$ by $p + q - k$. Show that

$$U_k(p, q, 1; x) - kxU_k(p, q-1, 1; x) = U_{k+1}(p, q, 1; x)$$

and, hence, that

$$U_k(p, q, 1; x) = T^{q-k-1}(T)_{k,x}(T)_{p,x}, \qquad T^k \equiv T_k(x)$$

with $(T)_{k,x} = T(T-x)\cdots(T-(k-1)x)$. Notice that the complement of a triangle of side $k$ in rectangle $p$ by $q$ is

$$U_{q-k}(p-k, q, 1; x) = T^{k-1}(T)_{q-k}(T)_{p-k}$$

with $(T)_k \equiv (T)_{k,x}$ for brevity, and

$$U_q(p, q, 1; x) = T^{-1}(T)_p(T)_q = R_{p,q}(x)$$

with $R_{p,q}(x)$ the rectangular polynomial. Comparing with Problem 29, notice the identity

$$S_n(x) = T(T+x)\cdots(T+(n-1)x)$$

$$= T(T-x)^2\cdots(T-(n-1)x)^2, \qquad T^k \equiv T_k(x)$$

31. (*a*) With $A_{nm}(t)$ the hit polynomial for a triangle of side $n - m$ in a square of side $n$, show that

$$A_{nm}(t) = \sum_{k=0} \binom{n-m}{k} (t-1)^k A_{n-k,m+1}(t)$$

(*b*) From the instance $m = 0$ of this, show that

$$\begin{aligned} A_1(t, u) &= \sum_{n=0} A_{n1} u^n / n! \\ &= e^{u(1-t)} \exp uA(t), \qquad A^n(t) \equiv A_{n0}(t) \\ &= (1-t)[e^{u(t-1)} - t]^{-1} \end{aligned}$$

(*c*) Write

$$A_m(t, u) = \sum_{n=0} A_{n+m-1,m}(t) u^n / n!, \qquad A_{m-1,m}(t) = (m-1)!$$

use (*a*) to show that

$$A_{m+1}(t, u) = e^{u(1-t)} \frac{d}{du} A_m(t, u), \qquad m > 0$$

and, hence, that

$$\begin{aligned} A_m(t, u) &= (m-1)!\,[A_1(t, u)]^m \\ &= (m-1) A_1(t, u) A_{m-1}(t, u) \end{aligned}$$

$$A_{n+m-1,m}(t) = (m-1) \sum_{k=0} \binom{n}{k} A_{n-k,1}(t) A_{k+m-2,m-1}(t)$$

32. For the hit polynomial $A_{21^{n-2}}(t)$ of Simon Newcomb's problem, write

$$A_{21^{n-2}}(t) = \sum_{r=0} A_{21^{n-2},r} t^r$$

and with

$$A_{n1}(t) = A_n(t) = \Sigma A_{n,r} t^r$$

show that

$$A_{21^{n-2},r} = A_{n,r} + A_{n-1,r} - A_{n-1,r-1}$$

33. In the notation of Problem 32, determine the following table for $a_{321^{n-5},r}$

| $n \backslash r$ | 1 | 2 | 3 | 4 | 5 | 6 |
|---|---|---|---|---|---|---|
| 5 | 1 | 6 | 3 | | | |
| 6 | 1 | 17 | 33 | 9 | | |
| 7 | 1 | 40 | 184 | 168 | 27 | |
| 8 | 1 | 87 | 792 | 1592 | 807 | 81 |

Note that the polynomials $A_{[s]}(t)$, $s = 321^{n-5}$, as functions of $n$ satisfy equation (50).

34. With $A_n(t)$ as in Problem 32, and $H_n(t) = (t-1)^{-n} A_n(t)$ as in Problem 2.2(*d*), so that

$$\exp uH(t) = (1-t)(e^u - t)^{-1}$$

and

$$(H+1)^n = tH_n(t) + (1-t)\delta_{n0}, \qquad H^k \equiv H_k(t)$$

show that, with $f(x)$ an arbitrary polynomial,

$$f(H+1) = tf(H) + (1-t)f(0), \qquad H^k \equiv H_k(t)$$

and, hence, that

$$\binom{H+1}{m} = t\binom{H}{m}, \qquad m > 0$$

$$\binom{H+r}{m} = t^r\binom{H}{m}, \qquad m \geq r$$

$$(t-1)^r\binom{H}{m} = \binom{H}{m-r}, \qquad m \geq r$$

$$(t-1)^m\binom{H}{m} = 1$$

35. Using the results of Problem 34 and

$$A_n(t) = (t-1)^n H_n(t) = \sum_{r=0}^{n-1} A_{nr}t^r$$

show that

$$H_n(t) = \sum_{r=0} A_{nr}\binom{H+r}{n}, \qquad H^k \equiv H_k(t)$$

and, hence, that

$$x^n = \sum_{r=0}^{n-1}\binom{x+r}{n}A_{nr} \qquad \text{(Worpitzky, 37)}$$

In particular,

$$x = \binom{x}{1}$$

$$x^2 = \binom{x}{2} + \binom{x+1}{2}$$

$$x^3 = \binom{x}{3} + 4\binom{x+1}{3} + \binom{x+2}{3}$$

36. For Simon Newcomb's problem with specification $(i^k)$

$$A_k^{(i)}(t) = A_{ik}(t) = \frac{(A)_i^k}{i!^k} = \binom{A}{i}^k = (t-1)^{ik}\binom{H}{i}^k$$

$$= \sum_{r=0}^{i(k-1)} A_{k,r}^{(i)}t^r$$

Derive the relation

$$\binom{H}{i}^k = \sum_{r=0} A_{k,r}^{(i)}\binom{H+r}{ik}$$

Hence,

$$\binom{x}{i}^k = \sum_{r=0} A_{k,r}^{(i)}\binom{x+r}{ik}$$

For $i = 1$, this reduces to the result of the preceding problem; so $A_{kr}^{(1)} = A_{kr}$.

37. Take $i = 2$ in Problem 36, and give $x$ the values 2, 3, $\cdots$, so that

$$1 = A^{(2)}_{n,2n-2}$$
$$3^n = A^{(2)}_{n,2n-3} + (2n+1)A^{(2)}_{n,2n-2}$$
$$6^n = A^{(2)}_{n,2n-4} + (2n+1)A^{(2)}_{n,2n-3} + \binom{2n+2}{2} A^{(2)}_{n,2n-2}$$

and so on. Show that

$$A^{(2)}_{n,2n-2-k} = \sum_{j=0}^{k} (-1)^j \binom{2n+1}{j} \binom{k+2-j}{2}^n$$

Verify the table for $A^2_{2n,r}$

| $n\backslash r$ | 0 | 1 | 2 | 3 | 4 | 5 | 6 |
|---|---|---|---|---|---|---|---|
| 1 | 1 | | | | | | |
| 2 | 1 | 4 | 1 | | | | |
| 3 | 1 | 20 | 48 | 20 | 1 | | |
| 4 | 1 | 72 | 603 | 1168 | 603 | 72 | 1 |

38. Show similarly that for general $i$

$$1 = A^{(i)}_{k,ik-i}$$
$$(i+1)^k = A^{(i)}_{k,ik-i-1} + (ik+1)A^{(i)}_{k,ik-1}$$
$$\binom{i+2}{2}^k = A^{(i)}_{k,ik-i-2} + (ik+1)A^{(i)}_{k,ik-i-1} + \binom{ik+2}{2} A^{(i)}_{k,ik-1}$$

and, hence, that

$$A^{(i)}_{k,ik-i-j} = \sum_{s=0}^{j} (-1)^s \binom{ik+1}{s} \binom{i+j-s}{i}^k \qquad \text{(Carlitz, } \mathbf{1}\text{)}$$

# Index

**Library of Congress Cataloging in Publication Data**

Riordan, John, 1903-
An introduction to combinatorial analysis.

Reprint of the ed. published by Wiley, New York, in series: A Wiley publication in mathematical statistics.
1. Combinatorial analysis. I. Title.
[QA164.R53 1980] 511'.6 80-337
ISBN 0-691-08262-6
ISBN 0-691-02365-4 pbk.